Full Stack Quarkus and React

Hands-on full stack web development with Java, React, and Kubernetes

Marc Nuri San Felix

BIRMINGHAM—MUMBAI

Full Stack Quarkus and React

Group Product Manager: Pavan Ramchandani

Publishing Product Manager: Kushal Dave

Senior Editor: Mark D'Souza

Senior Content Development Editor: Rashi Dubey

Technical Editor: Simran Udasi

Copy Editor: Safis Editing

Project Coordinator: Sonam Pandey

Proofreader: Safis Editing

Indexer: Hemangini Bari

Production Designer: Joshua Misquitta

Marketing Coordinator: Anamika Singh

First published: November 2022

Production reference: 1041122

Published by Packt Publishing Ltd.

Livery Place

35 Livery Street

Birmingham

B3 2PB, UK.

ISBN 978-1-80056-273-8

www.packt.com

To Julia, for her unconditional support on this journey. To Aitana and Álex, hoping that one day you'll read this book and it sparks your curiosity for computer science and technology.

– Marc Nuri San Felix

Foreword

I have known *Marc Nuri* for many years at Red Hat. As well as being an incredible person who is always willing to help when you've got a problem, Marc is always open to having a conversation about improving any piece of code he's writing.

At Red Hat, he is doing a great job of creating tools that improve the experience of Java developers transitioning to containers and Kubernetes technology. I can't think of anyone better qualified than Marc to teach you about Java and Kubernetes.

Marc is also an international speaker, and in his presentations he makes it clear why you should use something by including easy-to-understand examples that use a wide range of Java technologies, such as Maven/Gradle or Spring Boot/Quarkus.

In *Full Stack Quarkus and React*, you'll learn how to develop a Java backend application using Quarkus, the new Kubernetes-native Java framework that helps you develop supersonic (starting in milliseconds) and subatomic (minimizing its memory footprint) Java applications. Beyond the basics of Quarkus, you'll learn how to secure the backend – also known as making the application testable. As well as the backend, you'll develop the frontend using React, and integrate with Quarkus. Finally, you'll learn how to deploy the application in Kubernetes, the de-facto platform for releasing microservices, in just one step (no YAML) which is the developer's dream.

Get ready to become a Supersonic Subatomic Java developer with this book.

Alex Soto - Java Champion, Director of Developer Experience at Red Hat

Contributors

About the author

Marc started his career as a freelance web application developer, creating and maintaining software for the transportation/courier industry. A few years ago, Marc started working as an employee for different companies, where he helped build scalable web applications for different industries (retail, procurement software, e-commerce, etc.). He is now a professional open source developer focused on the creation and maintenance of developer tools for Java developers and Kubernetes.

I want to thank the Packt team and everyone else who helped out and worked on making this book come true; it's been quite a ride and a great team effort.

Special thanks to my family and everyone else who supported me and had the patience to be around me during this intense period.

About the reviewers

James Brooks is a full-stack developer with a wide range of professional experience. He's worked on a Windows desktop app and embedded devices at General Atomics, enterprise web-based applications at J.B. Hunt, and Medicare pricing and US Tax Court contracts at Flexion.

He's worked with a few Java frameworks and is a big fan of Quarkus. His day job sees him contributing to a React/Node application. You can check out his dev profile page at `https://james.oranbrooks.com`. After hours, he runs Hidden Gems Digital Marketing, a digital marketing agency based out of Fayetteville, AR, US. He has a passion for the craft of software development and approaches it from an engineering and quality mindset.

Marcus Melo is a software engineer passionate about delivering efficient and reliable solutions. He has a master's degree in computer science and certifications such as AWS Cloud Practitioner, a Java Programmer certification (SCJP 6), Scrum Master Certified, and AZ-900 Azure Fundamentals. He has about 16 years of experience in Java programming and, recently, experience in Java frameworks, some Quarkus, SpringBoot, and Angular, using some design patterns and working with agile methodology (Scrum). In addition, he worked as a developer at BB Tecnologia e Serviços, a division responsible for developing software solutions for one of the biggest banks in Brazil and many other companies. He currently works at Adentis in Portugal.

Wagner Ricardo Wagner is a software architect, trained in information systems. With more than 15 years of experience in the software market, he has participated in several projects in Brazil, Uruguay, and Paraguay, focused on the tax and financial sector. He has experience in the cloud, Java, Node, Go, databases, Google Cloud, K8s, DDD, clean code, architecture and code design, change data capture, NoSQL, and so on.

My thanks are directed to God, my family, and my friends, who always supported me throughout the journey. Also, I couldn't fail to mention Jean Pachla and Nícolas Tischler for always providing me with great challenges and opportunities for my career to grow.

Table of Contents

6

Building a Native Image   105

Part 2– Creating a Frontend with React

7

Bootstrapping the React Project   117

Part 3– Deploying Your Application to the Cloud

12

Deploying Your Application to Kubernetes   231

Preface

React has established itself as one of the most popular and widely adopted JavaScript libraries due to its simplistic yet scalable ability to create applications. Quarkus is a fantastic alternative for backend development by boosting developer productivity with features such as prebuilt integrations, application services, and more that bring a new revolutionary developer experience to Java.

This book provides hands-on experience to get started with creating and deploying an end-to-end web application with Quarkus and React. For a simpler understanding, the book has been divided into three main parts. In the first part, you will begin with a basic introduction to Quarkus and its features. You will be taught how to bootstrap a Quarkus project from scratch to create a secure, tested HTTP server for your backend. The second part focuses on the frontend; you will learn how to create a React project from scratch to build the application's user interface and how to integrate it with the Quarkus backend. The last part shows you how to create cluster configuration manifests and how to deploy them into Kubernetes, along with other alternatives such as Fly.io.

By the end of the book, you will be confident in your skills to combine the robustness of both frameworks to create and deploy standalone, fully functional web applications.

Who this book is for

This book is for backend web developers who have at least some basic experience with Java and who would like to learn React to build full stack apps, by integrating it with a Quarkus-based backend. It is also for frontend web developers with at least some basic experience with JavaScript who would like to implement a backend in Quarkus, integrate it with their frontend, and create full stack web applications.

Basic knowledge of Java and JavaScript is recommended, but any developer experienced just with Java or JavaScript should be able to follow the backend and frontend parts with no problem.

What this book covers

Chapter 1, *Bootstrapping the Project*, introduces Quarkus and the tooling that will be used, and shows you how to bootstrap a project for an application that will be implemented throughout the book.

Chapter 2, *Adding Persistence*, shows how to implement a persistence layer in Quarkus using Hibernate and the **Java Persistence API (JPA)**.

Chapter 3, *Creating the HTTP API*, explains how to create a service layer and use dependency injection to implement an HTTP API in Quarkus.

Chapter 4, Securing the Application, shows how to provide a security layer based on **JSON web tokens (JWTs)** to protect the HTTP API.

Chapter 5, Testing Your Backend, introduces the Quarkus testing framework and Quarkus Dev Services and explains how to use them to implement tests for your application.

Chapter 6, Building a Native Image, shows you how to create a native executable for your Quarkus application.

Chapter 7, Bootstrapping the React Project, introduces React and the libraries that will be used to implement the frontend part of the application, and shows you how to bootstrap the project.

Chapter 8, Creating the Login Page, explains how to implement the authorization infrastructure and how to create the login page for the frontend side of the application.

Chapter 9, Creating the Main Application, shows how to implement the core features of the application and how to consume the backend's HTTP API.

Chapter 10, Testing Your Frontend, explains how to implement unit and integration tests for your frontend applications based on React.

Chapter 11, Quarkus Integration, shows how to integrate the frontend application with Quarkus.

Chapter 12, Deploying Your Application to Kubernetes, gives a quick overview of Kubernetes and explains how to deploy the application to a Minikube Kubernetes cluster.

Chapter 13, Deploying Your Application to Fly.io, shows you how to deploy the application to Fly.io and make it publicly available.

Chapter 14, Creating a Continuous Integration Pipeline, introduces GitHub Actions and explains how to create a **continuous integration (CI)** pipeline.

To get the most out of this book

Basic knowledge of Java and JavaScript is recommended, but any developer experienced just with Java or JavaScript should be able to follow the complete book with no problem.

You will need the latest Java JDK LTS version, the latest Node.js LTS version, and a working Docker environment.

Software/hardware covered in the book	Operating system requirements
Quarkus 2	Windows, macOS, or Linux
React 18	

If you are using the digital version of this book, we advise you to type the code yourself or access the code from the book's GitHub repository (a link is available in the next section). Doing so will help you avoid any potential errors related to the copying and pasting of code.

Download the example code files

You can download the example code files for this book from GitHub at `https://github.com/PacktPublishing/Full-Stack-Quarkus-and-React`. If there's an update to the code, it will be updated in the GitHub repository.

We also have other code bundles from our rich catalog of books and videos available at `https://github.com/PacktPublishing/`. Check them out!

Download the color images

We also provide a PDF file that has color images of the screenshots and diagrams used in this book. You can download it here: `https://packt.link/yoqoD`.

Conventions used

There are a number of text conventions used throughout this book.

`Code in text`: Indicates code words in text, database table names, folder names, filenames, file extensions, pathnames, dummy URLs, user input, and Twitter handles. Here is an example: "The project includes the `.mvn` directory, and the `mvnw` and `mvnw.cmd` executable files."

A block of code is set as follows:

```
<dependency>
  <groupId>io.quarkus</groupId>
  <artifactId>quarkus-resteasy-reactive</artifactId>
</dependency>
```

Any command-line input or output is written as follows:

```
./mvnw quarkus:dev
```

Bold: Indicates a new term, an important word, or words that you see on screen. For instance, words in menus or dialog boxes appear in **bold**. Here is an example: "For this, we need to create a new debug configuration from the **Run | Edit Configurations** menu."

Tips or important notes
Appear like this.

Get in touch

Feedback from our readers is always welcome.

General feedback: If you have questions about any aspect of this book, email us at `customercare@packtpub.com` and mention the book title in the subject of your message.

Errata: Although we have taken every care to ensure the accuracy of our content, mistakes do happen. If you have found a mistake in this book, we would be grateful if you would report this to us. Please visit www.packtpub.com/support/errata and fill in the form.

Piracy: If you come across any illegal copies of our works in any form on the internet, we would be grateful if you would provide us with the location address or website name. Please contact us at `copyright@packt.com` with a link to the material.

If you are interested in becoming an author: If there is a topic that you have expertise in and you are interested in either writing or contributing to a book, please visit `authors.packtpub.com`.

Download a free PDF copy of this book

Thanks for purchasing this book!

Do you like to read on the go but are unable to carry your print books everywhere?

Is your eBook purchase not compatible with the device of your choice?

Don't worry, now with every Packt book you get a DRM-free PDF version of that book at no cost.

Read anywhere, any place, on any device. Search, copy, and paste code from your favorite technical books directly into your application.

The perks don't stop there, you can get exclusive access to discounts, newsletters, and great free content in your inbox daily!

Follow these simple steps to get the benefits:

1. Scan the QR code or visit the link below:

https://packt.link/free-ebook/9781800562738

2. Submit your proof of purchase
3. That's it! We'll send your free PDF and other benefits to your email directly

Part 1–
Creating a Backend
with Quarkus

This section focuses on the knowledge and skills required to implement a backend in Java by leveraging the Quarkus framework and its extensions. In this part, you will learn how to bootstrap a Quarkus project from scratch to create a secure, tested HTTP server for your backend.

In this part, we cover the following chapters:

- *Chapter 1, Bootstrapping the Project*
- *Chapter 2, Adding Persistence*
- *Chapter 3, Creating the HTTP API*
- *Chapter 4, Securing the Application*
- *Chapter 5, Testing Your Backend*
- *Chapter 6, Building a Native Image*

1
Bootstrapping the Project

In this book, we'll be creating a full-stack web application from scratch using **Quarkus** for the backend and **React** for the frontend. Quarkus is a new framework that aims to turn around the current **Java** ecosystem by simplifying the development of **cloud-native** applications and improving the developer experience. ReactJS is one of the most popular frontend JavaScript libraries. By the end of the book, you'll be able to combine both frameworks to create and deploy a full-featured task manager web application.

In this chapter, we'll provide a basic introduction to Quarkus and the tools that we'll be using across this section. Then, we'll create a Quarkus project from scratch and explain the basic architecture and structure of the application. By the end of this chapter, you should be able to create a Quarkus project and have a working development environment to implement the new features. You should also be able to package and run your application from your machine.

In this chapter, we're going to cover the following topics:

- What is Quarkus?
- Setting up the work environment with IntelliJ IDEA
- Bootstrapping a Quarkus application
- Project structure and dependencies

Technical requirements

You will need the latest Java JDK LTS version (at the time of writing, Java 17). In this book, we will be using Fedora Linux, but you can use Windows or macOS as well.

You can download the full source code for this chapter from `https://github.com/PacktPublishing/Full-Stack-Quarkus-and-React/tree/main/chapter-01`.

What is Quarkus?

Java has been around for more than 25 years now. It is one of the most used programming languages, especially for developing enterprise applications. Its rich ecosystem, extensive and open community, and *"write once, run anywhere"* approach, have made it the de facto choice for creating enterprise software for decades.

However, things are changing now. We are in the era of *the cloud*, *Kubernetes*, and *container images*. Things such as startup time or memory footprint, which were not that significant before, are becoming more relevant these days. Java is losing its pace compared to other languages, which are specifically tailored for these new environments.

Quarkus is a new Java framework that was first released in 2019. It provides similar features to those of other mainstream Java frameworks such as Spring Boot or Micronaut. The main goals of Quarkus are to improve the application's startup time, its memory footprint, and the developer experience.

Quarkus was built from the ground up to be *cloud-native* from the start. Regardless of how you package your application, boot time, and memory consumption, Quarkus performs much better than the alternatives. That's why Quarkus is also known as *Supersonic Subatomic Java*.

Quarkus is built on top of long-lived proven **standards** and **libraries**. When Quarkus was designed, the team decided to rely on existing tools, frameworks, and standards instead of building something new from scratch. This was done so that developers don't have to spend time learning something new and can focus on building more performant applications instead, taking advantage of their experience.

Quarkus uses the **Java Enterprise Edition (EE) Contexts and Dependency Injection (CDI)** standard for dependency injection; **MicroProfile**, a community-driven specification to optimize Java EE for a microservice architecture and for configuration and monitoring; **Java Persistence API (JPA)** annotations to define the **object-relational mapping (ORM)**; **Jakarta RESTful Web Services (JAX-RS)** annotations to define the REST controllers; and many more technologies in a long, ever-growing list.

Quarkus brings a new **developer experience** to Java. One of the main drawbacks of Java when compared to other languages is the traditionally slow development cycles. A simple change to a line of code usually involves recompiling, repackaging, and restarting the application. This process could take anywhere from seconds to minutes, completely reducing the developer's productivity. Quarkus aims at fixing this pain point by providing live coding, a unified configuration, a developer UI, and many more tools to bring joy back to developers.

Quarkus combines the traditional **imperative** coding style with the cloud-native friendly **reactive** coding style approach. Regardless of the type of application you are building, Quarkus provides first-class support for both paradigms. The cloud has brought new architectures to our systems, whether microservices, serverless, or event-driven. In this book, we'll explore the new reactive non-blocking style, which will bring massive performance improvements when compared to the more classic imperative approach.

Quarkus is a free open source project that was initially released in March 2019. It is not only a fast-growing project in terms of adoption, but its community is soaring too. Despite its youth, it already has around 600 contributors (at the time of writing) and a growing ecosystem of extensions (`https://github.com/quarkiverse`). There's already an extensive list of publications and bibliography available too.

Quarkus makes it extremely easy to create **native** executables for your platform. The framework provides almost transparent integration with *GraalVM*, a high-performance **Java Development Kit** (**JDK**) distribution that allows you to compile your Java code into a standalone binary executable.

Setting up the work environment with IntelliJ IDEA

In this book, we will develop a full-stack web application with Quarkus and React. Thus, we need to get started with writing the code for our application. We will be using **IntelliJ IDEA** for this purpose; however, you are free to use an **integrated development environment** (IDE) or editor of your choice. IntelliJ IDEA is a good choice because its ultimate version works well for both frontend and backend development.

IntelliJ IDEA is an IDE developed by *JetBrains*, which is based on the IntelliJ open source platform. The product is available in two flavors: the free community edition and the Ultimate Edition, which requires a paid subscription. However, there is an early access program that gives you free access to the early builds of the Ultimate version.

The easiest way to get IntelliJ IDEA is to go to the product page and download the appropriate bundle for your system: `https://www.jetbrains.com/idea/download`.

Bootstrapping a Quarkus application

Quarkus provides several options for bootstrapping an application. The easiest method involves using the `https://code.quarkus.io` page. You can also generate a new application by using the provided **command-line interface** (**CLI**) tool or even by running a **Maven** goal. We'll be using the web interface since it's the simplest approach and has a very straightforward wizard to customize the options and extension dependencies. However, you are free to explore other alternatives.

> **Maven goal**
>
> Maven is a build automation and project management tool mainly used in Java. A goal is a specific task that you can perform through Maven. Tasks such as compiling, packaging, and testing in Maven projects are performed via goal executions.

We are going to generate a Java 17 project with Maven and the latest available Quarkus version. Quarkus provides a web-based tool that allows you to create and customize an initial project. Let us take advantage of that by going through the following steps:

1. Navigate with your web browser to `https://code.quarkus.io`.

   You'll be greeted with the following wizard:

Figure 1.1 – A screenshot of the code.quarkus.io web interface

2. In the **Group** field, set the Maven project Group ID.

   The generated project will include a package named with this value containing some sample starter code. I'm going to use `com.example.fullstack`, but you can use any other value.

3. In the **Artifact** field, define the Maven artifact ID. In my case, I will use `reactive`.

4. Select the base dependencies for our project. For now, we are only going to add a dependency to enable the implementation of HTTP endpoints. Since we want to take full advantage of Quarkus' reactive features, we'll use **RESTEasy Reactive**. You can filter through the dependencies by typing into the **Filters** text field and then checking the required dependency checkbox:

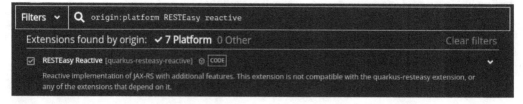

Figure 1.2 – A screenshot of the dependency selection section in the code.quarkus.io web interface

In the following chapters, we'll be including more dependencies as we add functionality to the project.

5.   Click on the **Generate your app** button and download a ZIP package with the bootstrapped project.

Now that you have generated your Quarkus application, let us learn about the structure of the project and what features each of the project's initial dependencies and plugins provides.

Project structure and dependencies

To start exploring the project's structure, let us go through the following steps:

1.   Extract the generated ZIP package from the preceding bootstrapping exercise to the project's definitive location. I extracted the ZIP to my dedicated directory for projects – you can extract it wherever you prefer but make sure you remember this location since this is the project we will be working on throughout the book.

2.   Open the project in IntelliJ IDEA or the IDE of your choice.

IntelliJ should automatically detect your Maven project and load its dependencies. In case it doesn't, you can perform this step manually by right-clicking the pom.xml file and clicking on the **Add as Maven Project** menu item:

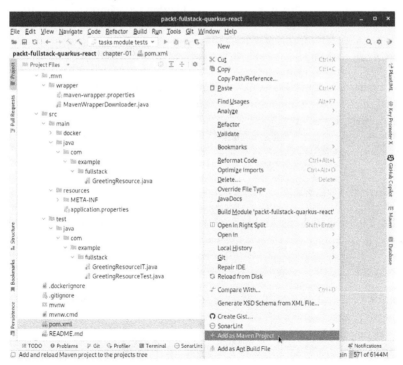

Figure 1.3 – A screenshot of IntelliJ IDEA and the Add as Maven Project context menu

Let's now explore the content and structure of the project and the provided sample code.

Maven Wrapper

The project includes a **Maven Wrapper** setup. Maven Wrapper is a tool that allows project users to run a consistent version of Maven across different build environments. The tool also allows you to run Maven without the need to have a global Maven installation. The project includes the .mvn directory and the mvnw and mvnw.cmd executable files.

You should be able to invoke Maven goals from a terminal in your project root. If you are in a Linux or macOS environment, you should be able to execute the wrapper by running ./mvnw. If you are on Windows, you can execute the Wrapper by running ./mvnw from a **PowerShell** terminal, or mvnw from a standard **cmd.exe** terminal. Now that we've seen the provided Maven Wrapper setup, let's focus, examine the Maven project configuration more closely, and analyze what each section accomplishes.

Maven project (pom.xml)

The Maven project is defined in the **Project Object Model** pom.xml file. This **XML** file is the main unit of work for Maven and collects all the information and configuration details that will be used by Maven to build the project.

Let's examine some of the sections of the pom.xml file that were bootstrapped for us from the Quarkus website.

Maven coordinates (GAV)

The **Maven coordinates**, also known as GAV, are the minimum required references for a project. These are the groupId, artifactId, and version fields that we defined in the web-based wizard when we bootstrapped the project:

```
<groupId>com.example.fullstack</groupId>
<artifactId>reactive</artifactId>
<version>1.0.0-SNAPSHOT</version>
```

These fields act as a unique identifier for the project and enable you to reference it in other projects just like a coordinate system.

Maven properties

The project comes with a set of predefined properties in an XML <properties> block. The following are the most important:

- maven.compiler.release

  ```
  <maven.compiler.release>17</maven.compiler.release>
  ```

This property sets the Java version for the project. In this case, both the sources and target classes will require a Java 17 version. This property is used by the Maven Compiler Plugin, and it was introduced in version 3.6 of the plugin. This property relies on the other `compiler-plugin.version` property, which you shouldn't change – or at least make sure it's always later than 3.6.

- `quarkus.platform.version`

```
<quarkus.platform.version>2.10.2.Final</quarkus.platform.
version>
```

This property specifies the Quarkus version in use. Whenever a new Quarkus version is released, this is the property that you should update to upgrade your project. For patch versions and non-breaking releases, this change should be enough. For other version updates, you might need to change some parts of your code too.

Dependency management

The `pom.xml` file contains a **dependency management** block with the following content:

```
<dependencyManagement>
  <dependencies>
    <dependency>
      <groupId>${quarkus.platform.group-id}</groupId>
      <artifactId>${quarkus.platform.artifact-id}
      </artifactId>
      <version>${quarkus.platform.version}</version>
      <type>pom</type>
      <scope>import</scope>
    </dependency>
  </dependencies>
</dependencyManagement>
```

This definition is important to set the version of the Quarkus extension dependencies. It's using placeholders for the following Maven properties found in the `properties` section to reference the effective dependency:

```
<quarkus.platform.artifact-id>quarkus-bom
</quarkus.platform.artifact-id>
<quarkus.platform.group-id>io.quarkus.platform
</quarkus.platform.group-id>
```

Under the hood, Maven is copying the dependency management section of the `io.quarkus.platform:quarkus-bom`, Quarkus' **Bill of Materials** (**BOM**), artifact to the current project. This process enforces the use of a consistent version for all of the provided Quarkus extensions that we'll see in the next section, *Dependencies*.

Dependencies

The following block in the project object model is the **dependencies** definition. These are the actual library dependencies of our project. Let's see what each of the bootstrapped dependencies does.

RESTEasy Reactive

In this book, we are going to explore the new reactive capabilities of Quarkus. RESTEasy Reactive is a Quarkus-specific implementation of the **JAX-RS** specification based on **Vert.x**. It takes full advantage of Quarkus' reactive non-blocking capabilities, which improve the overall application performance. The following code snippet defines the dependency for this library:

```
<dependency>
    <groupId>io.quarkus</groupId>
    <artifactId>quarkus-resteasy-reactive</artifactId>
</dependency>
```

JAX-RS is a Java EE or Jakarta EE API specification that enables the implementation of REST web services. It provides common annotations such as `@Path`, `@GET`, and `@POST`, which can be used to annotate classes and methods to implement HTTP endpoints. If you've dealt with J2EE, Java EE, or Jakarta EE before, you might already be familiar with these annotations.

This highlights one of the main advantages of Quarkus. The learning curve is very gentle since most of it is based on proven community standards and libraries.

Quarkus ArC

Quarkus ArC is the dependency injection solution provided by Quarkus. It is based on the Java EE CDI 2.0 specification – again, a proven, long-lived standard. The following code snippet specifies this dependency:

```
<dependency>
    <groupId>io.quarkus</groupId>
    <artifactId>quarkus-arc</artifactId>
</dependency>
```

One of the advantages of ArC, and most Quarkus extensions in general, is that it's build-time oriented. Most analysis and optimizations happen at build time, so none of this processing needs to be performed during the application startup. The result is an application that starts up nearly instantly.

Quarkus JUnit5

Quarkus JUnit5 is the main dependency for the Quarkus testing framework. It provides the @ QuarkusTest annotation, which is the main entry point for the test framework. The next code snippet configures this dependency:

```
<dependency>
    <groupId>io.quarkus</groupId>
    <artifactId>quarkus-junit5</artifactId>
    <scope>test</scope>
</dependency>
```

We'll examine this dependency and its features in more detail in *Chapter 5, Testing Your Backend*.

Rest Assured

Rest Assured is the last test dependency that was bootstrapped in the project. Although it's not provided by Quarkus, it's the recommended way to test its endpoints. The following code snippet is used to define this dependency; notice the groupId value is not io.quarkus anymore:

```
<dependency>
    <groupId>io.rest-assured</groupId>
    <artifactId>rest-assured</artifactId>
    <scope>test</scope>
</dependency>
```

We'll be using it to create the integration tests for our application.

Plugins

Along with the more common Maven plugins, the build plugins section contains an entry for the **Quarkus Maven plugin**. This plugin provides Maven goals for most of the Quarkus features. Whenever we invoke any Maven command with a quarkus: prefix, this is the plugin that will be responsible for the execution.

Profiles

The last section in the pom.xml file is the one dedicated to profiles. The bootstrapped project contains a single profile with the native ID. We can activate this profile either by using the Maven profile selection flag, -Pnative, or by providing a -Dnative system property (see the activation configuration):

```
88        <profiles>
89          <profile>
90            <id>native</id>
91            <activation>
92              <property>
93                <name>native</name>
94              </property>
95            </activation>
```

Figure 1.4 – A screenshot of the beginning of the profiles section in pom.xml

The profile provides some specific configurations to run tests that partially override the one provided in the build or plugins section. However, the most important part is the quarkus.package.type property. This is the property that instructs Quarkus to build a native binary for our platform. When we package our application with this profile (./mvnw clean package -Pnative), we'll get a binary file instead of a standard **Java archive (JAR)** package.

We'll explore profiles in more detail in *Chapter 6, Building a Native Image*.

Source files

The bootstrapped project has the regular Java project structure. In addition to the pom.xml project file in the root directory, you will find a src subdirectory that contains the project sources.

Application properties

The application.properties file is located in the src/main/resources directory. This file contains the main configuration for our project. We'll be modifying the application configuration and behavior by adding entries to it.

Under the hood, Quarkus uses **SmallRye Config**, which is an implementation of the **Eclipse MicroProfile Configuration feature** spec. This is another of the battle-tested standards on which Quarkus is based.

This is a standard property file. Each entry is added in a new line. For each line, the config key and the config value are separated by an = sign.

For example, the code to set the application server port would be as follows:

```
quarkus.http.port=8082
```

The application.properties file can also be used to define values that can be injected into your application.

Let's say you defined the following property:

```
publisher.name=Packt
```

You could inject the preceding property into your application using the following snippet:

```
@ConfigProperty(name = "publisher.name")
String publisherName;
```

Profiles

Quarkus provides the option to build and execute the application based on different **profile** configurations. Depending on the target **environment**, you might want to select a specific profile that provides a valid configuration for that environment.

Quarkus has three profiles – `dev`, which is activated in development mode, `test`, which is activated when the tests are executed, and `prod`, which is the default profile when the others don't apply.

The same file is used for all profiles; to configure an option for a specific profile, you need to prefix the configuration key with `%` and the profile name, except for `prod`, which is the default profile and doesn't require a prefix.

For the previous server port example, we can set the server port in `dev` mode as follows:

```
%dev.quarkus.http.port=8082
```

In general, we'll be adding configuration for `prod`, and provide overrides for `dev` mode when needed.

Static resources

The project contains an `index.html` file in the `src/main/resources/META-INF/resources` directory. This file will be automatically served from the underlying application's HTTP server. When pointing a browser to the root path of the application (`http://localhost:8080`), you will be greeted with this landing page that was bootstrapped for us:

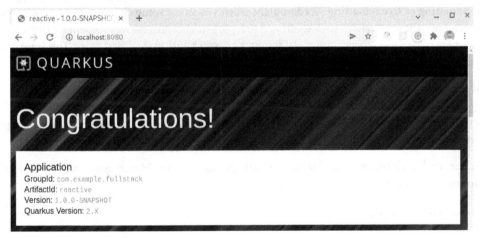

Figure 1.5 – A screenshot of a browser pointing to http://localhost:8080

By default, Quarkus will serve any static file that is located in this directory. However, for our application, we'll be using an alternative method since this approach is not compatible with frontend routing. In *Chapter 11*, *Quarkus Integration*, I'll show you how to implement an **API gateway** that will be used as an alternative to serving the static resources.

Java code

The bootstrapped project contains some sample code. A `GreetingResource` class is located in the standard `src/main/java` directory under the `com.example.fullstack` package. You will also find two tests for this class in the `src/main/test` directory under the same package: `GreetingResourceTest` and `GreetingResourceIT`. We will place the new code that we implement in the same root package grouped by features.

Docker files

The project contains some example **Docker** files in `src/main/docker`. These files can be used to create **container images** for your application. In *Chapter 12*, *Deploying Your Application to Kubernetes*, I'll show you how to create container images for the application. However, we'll be using **Eclipse JKube**, which requires a simpler configuration and doesn't need these Docker files. JKube is a Maven plugin that generates all of the required configurations for your application to be able to deploy it to Kubernetes; for this reason, it's not necessary to keep extra configuration files such as Docker or Kubernetes YAML files.

Now that we've seen the files and directory structure of the bootstrapped project, let us see how to perform the basic tasks that we will need to develop new features and deploy and run the application.

Development mode

For years, one of the main pain points for Java developers has been the lack of or very little support for **hot reloading** or **live reloading**. Traditionally, when you made some changes to your code, you had to recompile the application, package it, and redeploy it. This process was something that could take anywhere from a few seconds to several minutes or even hours in the worst cases. This is usually one of the disadvantages cited when people compare Java to other programming languages.

One of Quarkus' main goals is bringing joy back to developers, so, naturally, this was one of the priority points to address. Quarkus **development mode** runs your application and monitors your code. Whenever you change any of the Java application source or resource files, Quarkus detects these changes and performs a hot deployment. You just need to refresh your browser for the changes to take effect.

We can start the development mode by running the `quarkus:dev` Maven goal from the project's root directory as follows:

```
./mvnw quarkus:dev
```

You will see the following result:

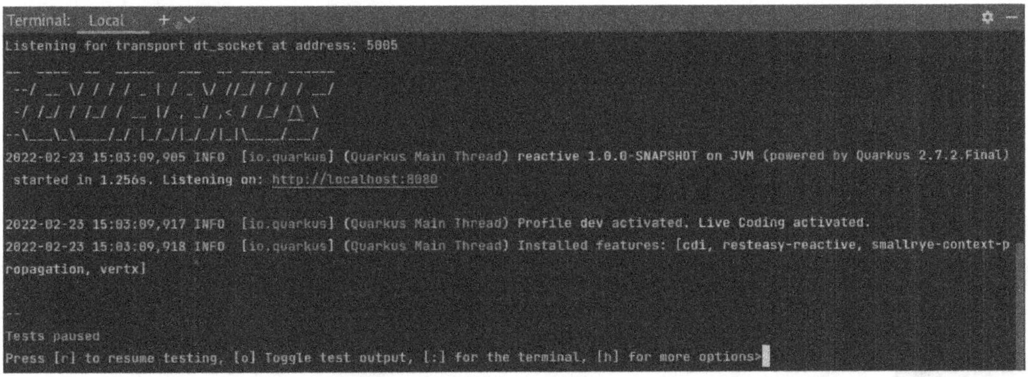

Figure 1.6 – A screenshot of the IntelliJ terminal running Quarkus development mode

If you check the preceding messages, you'll notice that Quarkus automatically selected for us the `dev` profile and started the live coding mode.

We can now point our browser to the URL of the sample endpoint that was bootstrapped (`http://localhost:8080/hello`). If everything went well, the browser will show the **Hello RESTEasy Reactive** message:

Figure 1.7 – A screenshot of the browser pointing to the hello endpoint

If we open the `GreetingResource` class in our IDE, we should be able to see the definition for this endpoint. We can change the greeting message to something else:

```
@GET
@Produces(MediaType.TEXT_PLAIN)
public String hello() {
    return "Hello Quarkus live coding!";
}
```

In the traditional Java world, we would now need to recompile and redeploy the application to be able to see the changes. However, if we reload the browser window, our modified message should be visible.

Debugging in development mode

If you check the messages in *Figure 1.6* closely, you'll notice that Quarkus has also enabled a remote debugging port:

```
Listening for transport dt_socket at address: 5005
```

This means we can easily start a debug session from IntelliJ IDEA. For this, we need to create a new debug configuration from the **Run** > **Edit Configurations…** menu:

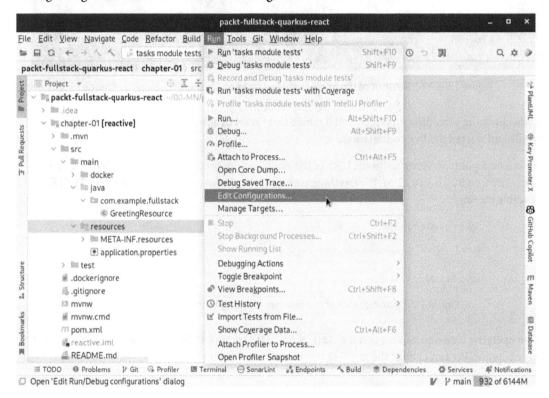

Figure 1.8 – A screenshot of the IntelliJ IDEA Run menu

From the **Run/Debug Configurations** screen, we need to create a new **Remote JVM Debug** configuration. The default options should be fine for Quarkus, so we only need to specify a name for this configuration:

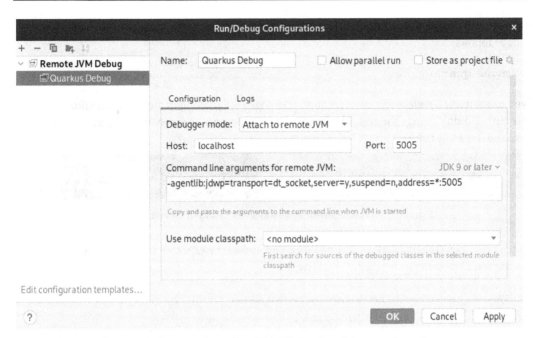

Figure 1.9 – A screenshot of IntelliJ IDEA Quarkus debug configuration

Once we save the configuration, we can run it and should be able to set a breakpoint on our endpoint definition. If we reload the browser window, the debugger should stop at our breakpoint.

When combined with easy debugging, live reloading is very powerful and will certainly improve our developer performance. Now that we know how to use the Quarkus development mode to implement and debug our code, let us see how to run the tests for our application.

Continuous testing

One of Quarkus 2.X's features is its ability to run tests continuously. This is a feature borrowed from other programming languages, such as *Ruby*, that have offered it for a long time. It is also a further step in achieving Quarkus's goal to bring back developer joy to Java.

For users who practice **test-driven development** (TDD), this will massively improve their development cycle performance. In a usual TDD process, developers first write a test for a feature and then implement the code that will make that test pass. This process is repeated for each of the properties of the feature and for each code refactor. Continuous test execution provides instant feedback and allows the developer to concentrate and focus on the implementation and not on the process.

When Quarkus is run in **continuous testing** mode, it will detect code changes to both code and tests. For each change it detects, it will re-run the relevant tests for the affected code.

Just as with the development mode, we can run a Maven command to start the continuous testing mode as follows:

```
./mvnw quarkus:test
```

If you recall, in the *Development mode* section, we changed the greeting in the GreetingResource class but we didn't change the test. The first thing we'll see once we invoke the quarkus:test Maven goal is a test failure:

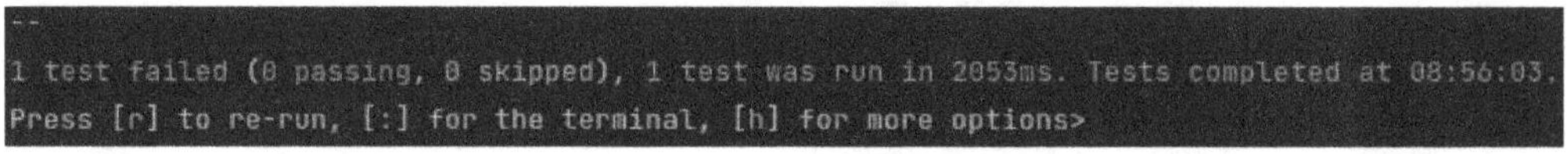

1 test failed (0 passing, 0 skipped), 1 test was run in 2053ms. Tests completed at 08:56:03. Press [r] to re-run, [:] for the terminal, [h] for more options>

Figure 1.10 – A screenshot of quarkus:test failing to be invoked

We can now open GreetingResourceTest and update the expected response body to the new greeting, Hello Quarkus live coding!:

```
@Test
public void testHelloEndpoint() {
    given()
      .when().get("/hello")
      .then()
         .statusCode(200)
         .body(is("Hello Quarkus live coding!"));
}
```

If we save the changes, the test should automatically re-run and it will be green again:

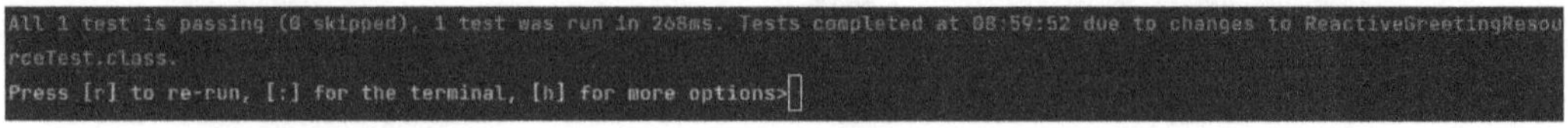

All 1 test is passing (0 skipped), 1 test was run in 268ms. Tests completed at 08:59:52 due to changes to ReactiveGreetingResourceTest.class.
Press [r] to re-run, [:] for the terminal, [h] for more options>

Figure 1.11 – A screenshot of quarkus:test passing invocation

Expanding on the TDD use case, if there was a new requirement to expose a /hello/world endpoint, the first step would be to add a new test:

```
@Test
public void testHelloWorldEndpoint() {
    given()
      .when().get("/hello/world")
      .then()
```

```
        .statusCode(200)
        .body(is("Hello world!")));
}
```

There is no implementation yet, so the execution would fail. We could then implement the new endpoint to make the test pass as follows:

```
@GET
@Produces(MediaType.TEXT_PLAIN)
@Path("/world")
public String helloWorld() {
    return "Hello world!";
}
```

Once tests pass, the next step could be to retrieve the endpoint value from an external service. So, we would modify the test, then the implementation, and start the cycle again. It should be clear now how the experience of the overall process is notably improved by continuous testing.

TDD ensures that the features defined in the provided specs are working using unit tests. This allows you to write code with fewer bugs and spend less time on long debugging sessions trying to fix errors. Now that we've seen how to perform TDD in Quarkus, let us see how to package the application for its distribution.

Packaging the application

The final step to being able to distribute and run the application would be to package it. Besides the native mode, which we already analyzed in the *Profiles* section, Quarkus offers the following package types:

- **fast-jar**: This is the default packaging mode. It creates a highly optimized runner JAR package, along with a directory and its dependencies.

- **uber-jar**: This mode will generate a fat JAR containing all of the required dependencies. This JAR package is suitable for distribution of the application on its own.

- **native**: This mode uses GraalVM to package your application into a single native binary executable file for your platform.

- **native-sources**: This type is intended for advanced users. It generates the files that will be needed by GraalVM to create the native image binary. It's like the native packaging type but stops before triggering the actual GraalVM invocation. This allows performing the GraalVM invocation in a separate step, which might be useful for CI/CD pipelines.

You can control the packaging mode by setting the `quarkus.package.type` Maven property. You can set this property in the `pom.xml` properties section or via the command line when running the Maven commands:

```
./mvnw -D"quarkus.package.type=uber-jar" clean package
```

For the moment, we'll be using the default packaging mode. You can package the application running the following command:

```
./mvnw clean package
```

If everything went well, you should now be able to run the application by executing the following:

```
java -jar target/quarkus-app/quarkus-run.jar
```

You should now be able to navigate to `http://localhost:8080` or any of the HTTP endpoints we created in the previous steps.

Summary

In this chapter, we gave a quick introduction to Quarkus and saw what its main advantages and breakthroughs for the Java ecosystem are. We created a new Quarkus project using the Quarkus website and opened it in IntelliJ IDEA. We reviewed the contents and structure of the bootstrapped project. We checked the dependencies, plugins, and the provided source code, and we learned how to package and run the sample application.

You should now have a broad view of Quarkus and be ready to start implementing the full-stack web application that we'll be developing throughout the book. In the next chapter, we'll see how to add a persistence layer to the task manager application, define the entity classes, and connect it to a database.

Questions

1. What is Quarkus?
2. How do you create a Quarkus project from scratch?
3. How do you run a Quarkus project in development mode?
4. What is TDD?
5. How do you package a Quarkus project?

2
Adding Persistence

In this chapter, we'll add a persistence layer to the application and learn how to connect it to a database. We will also define the **Entity** classes that will support the main functionality for the task manager web application we are building throughout the book. By the end of this chapter, you should be able to persist and store the data of your application in a database. You should also have a basic understanding of **Hibernate** and be able to define your own entities.

We will be covering the following topics in this chapter:

- Data persistence in Quarkus
- Implementing the task manager data model
- Quarkus Dev Services

Technical requirements

You will need the latest Java JDK LTS version (at the time of writing, Java 17). In this book, we will be using Fedora Linux, but you can use Windows or macOS as well.

You will need a working Docker environment to take advantage of Quarkus **Dev Services**. There are Docker packages available for most Linux distributions. If you are on a Windows or macOS machine, you can install **Docker Desktop**.

You can download the full source code for this chapter from `https://github.com/PacktPublishing/Full-Stack-Quarkus-and-React/tree/main/chapter-02`.

Data persistence in Quarkus

Data persistence is the means by which your application will be able to store and retrieve its data from one execution to the next. This means that any data that was input will survive after the process that created it has ended. Quarkus has a growing list of extensions to support data persistence. In *Chapter 1*, *Bootstrapping the Project*, we learned about Quarkus' support for both the imperative and reactive paradigms. Since the reactive approach will bring considerable performance improvements to our application, in this book, we're going to use **Hibernate Reactive**, a reactive persistence API, in its simplified **Panache** version, to provide data persistence to our application. We are also going to use a **PostgreSQL** database since there is a reactive client extension for Quarkus too. Let us now see which dependencies we should add to the project.

Adding dependencies to our project

We will start by adding all of the required dependencies for this chapter and then we'll evaluate what each of the dependencies provides. You can execute the following command in the project root and the *Quarkus Maven plugin* will add the dependencies for you:

```
./mvnw quarkus:add-extension -Dextensions=hibernate-reactive-
panache,reactive-pg-client,io.quarkus:quarkus-elytron-security-
common
```

Once executed, you should see the following message:

```
[INFO] --- quarkus-maven-plugin:2.7.2.Final:add-extension (default-cli) @ reactive ---
[INFO] [SUCCESS] ✅ Extension io.quarkus:quarkus-elytron-security-common has been installed
[INFO] [SUCCESS] ✅ Extension io.quarkus:quarkus-hibernate-reactive-panache has been installed
[INFO] [SUCCESS] ✅ Extension io.quarkus:quarkus-reactive-pg-client has been installed
```

Figure 2.1 – A screenshot of the execution result of the quarkus:add-extension command

Let us see what each of these dependencies provides.

Hibernate Reactive Panache

After executing the ./mvnw quarkus:add-extension command, the following dependency should be visible in our pom.xml:

```
<dependency>
  <groupId>io.quarkus</groupId>
  <artifactId>quarkus-hibernate-reactive-panache
  </artifactId>
</dependency>
```

Hibernate is one of the most mature **object-relational mapping (ORM)** libraries for the Java programming language. It was initially released in 2001 and is one of the most widely used frameworks for data persistence. It also provides an implementation for the **Java Persistence API (JPA)** specification.

ORM

ORM is a software engineering technique that enables applications to interact with a database in a platform-agnostic way using an object-oriented approach. Instead of writing database vendor-specific queries using **Structured Query Language (SQL)** or other languages, developers can interact with the database and define the entities using a more natural Java programming language.

JPA

JPA is the most widely adopted ORM specification for the Java programming language. There are multiple implementations and tooling available for most of the mainstream frameworks.

Panache is a Quarkus-specific, simplified version of Hibernate. It allows writing JPA entities without the need for repetitive boilerplate code. It also provides methods to perform **create, read, update, and delete (CRUD)** operations, besides defining and executing basic queries for your entity.

In this case, we are adding the reactive-flavored version of the dependency, which will allow us to perform non-blocking, asynchronous operations.

Reactive PostgreSQL Client

The previous execution of the `./mvnw quarkus:add-extension` command should add the following dependency to the `pom.xml` file:

```
<dependency>
    <groupId>io.quarkus</groupId>
    <artifactId>quarkus-reactive-pg-client</artifactId>
</dependency>
```

The **Reactive PostgreSQL Client** is a Quarkus-specific dependency that allows your application to connect to a PostgreSQL database using the reactive pattern. It's especially suited for Hibernate Reactive and RESTEasy Reactive since it allows you to take full advantage of the reactive, non-blocking approach, from the client's request to effective data retrieval and processing from the database.

Elytron Security Common

The previous execution of the `./mvnw quarkus:add-extension` command adds a third dependency to the `pom.xml` project configuration too:

```
<dependency>
    <groupId>io.quarkus</groupId>
```

```
    <artifactId>quarkus-elytron-security-common</artifactId>
</dependency>
```

This dependency is not related to data persistence; it provides some security tooling that we will use to encrypt the passwords. Since this dependency is directly maintained in the Quarkus project, it's a good choice for the needs of our application.

Now that we've added all of the required dependencies for this section, let us see how to configure Quarkus to take advantage of them.

Configuring Quarkus

To be able to start using Hibernate Reactive in our application, we need to add some basic configurations. In the *Application properties* section of *Chapter 1*, *Bootstrapping the Project*, we saw that the application configuration was defined in the `application.properties` file. This file should be empty at the moment; we need to add the following entries to configure Hibernate and PostgreSQL:

```
quarkus.datasource.db-kind=postgresql
quarkus.hibernate-orm.sql-load-script=import.sql
%dev.quarkus.hibernate-orm.database.generation=drop-
and-   create
%dev.quarkus.hibernate-orm.sql-load-script=import-dev.sql
%dev.quarkus.hibernate-orm.log.sql=true
```

Let us see what each of these configurations does:

- `quarkus.datasource.db-kind`

  This entry defines the type of database we'll be connecting to; other possible values could be h2, derby, mariadb, and so on. In our case, since we'll be using PostgreSQL as our database, the appropriate value is postgresql.

- `quarkus.hibernate-orm.sql-load-script`

  This property defines a path to a file containing SQL statements that will be executed when the application starts. By default, when running Quarkus in *dev* mode, it will try to automatically load a file with the name import.sql. However, since we also want to provide some initial data for *production*, we manually define the property for both environments. In this case, we've added an extra entry for the dev profile: %dev.quarkus.hibernate-orm.sql-load-script=import-dev.sql. When we launch the application in *dev* mode, the import-dev.sql file will be loaded; when we launch the application in *production* mode, the import.sql file will be loaded instead.

- `%dev.quarkus.hibernate-orm.database.generation`

  This property instructs Hibernate on what to do with the application schema upon startup. The default is for Hibernate to perform no action; however, in our case, we'll drop and create the schema each time we start the application in *dev* mode. Feel free to change this value if you want to persist example data between development iterations.

- `%dev.quarkus.hibernate-orm.log.sql`

  This entry configures Quarkus to log the SQL commands sent to the database in the console. This allows us to see how Hibernate transforms our Java code into SQL. It will be useful for this chapter since we are only defining the data model; it will allow us to see how the tables that our entities represent are created in the database. Feel free to enable and disable this property at your will if it turns out to be distracting at some point.

Now that we've seen how to set up and configure a data persistence layer with Hibernate Reactive, let us implement the data layer of our application.

Implementing the task manager data model

In this book, we are going to implement a full-featured task manager web application. In this chapter, we'll start by implementing the data model that we'll be using to persist the application's state in a database.

The following diagram shows the different classes/entities that we are going to implement and their relationships:

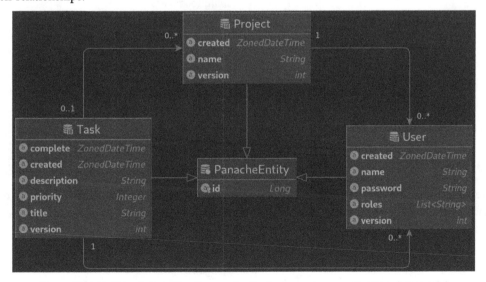

Figure 2.2 – Entity-relationship diagram for the task manager application data model

The model is very straightforward: the main entity is a `Task` that might be assigned to a `Project` and always belongs to a `User`. The application is quite simple: a user logs into the task manager and can create different tasks. Each task has a title and a longer optional description. The user may assign a priority to the task and mark it as complete when it's done. In addition, the user can create different projects that can be used to optionally group the tasks. In this section, I'll explain each entity in detail; however, let's start by removing some of the example files that were created by the `code.quarkus.io` project generator.

Deleting the bootstrapped example classes and files

Before implementing the required classes, we'll start by removing the example source code provided by the `code.quarkus.io` project initializer. We'll delete the following classes and files from the project:

- `src/main/java/com/example/fullstack/GreetingResource.java`

- `src/test/java/com/example/fullstack/GreetingResourceTest.java`

- `src/test/java/com/example/fullstack/GreetingResourceIT.java`

- `src/main/resources/META-INF/resources/index.html`

To delete these files, you can right-click on them and press on the **Delete…** menu item.

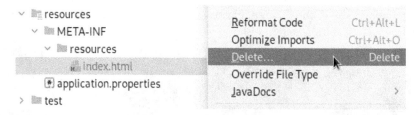

Figure 2.3 – A screenshot of IntelliJ IDEA's Delete… context menu entry

Let us now create the data model entities.

Creating the task manager entities

We are going to implement three entities for the task manager: `Task`, `Project`, and `User`. All of them will extend the `PanacheEntity` class. Let us now see what features the Panache entity provides.

Panache repositories and entities

In the *Adding dependencies* section, we learned that Panache is a Quarkus-specific extension that allows us to write JPA entities with less boilerplate code. The Panache extension allows us to implement our persistence layer using either the **repository pattern** or the **active record pattern**. Depending on the development pattern we select, the extension provides us with two different base classes: `PanacheEntity`, which will be used for the active record pattern, and `PanacheRepository`, which will be used when dealing with the repository pattern.

The repository pattern was first introduced as part of **domain-driven design** (**DDD**). It defines a repository as a class or component that encapsulates the logic required to access, create, update, or delete the entity objects or their aggregates. Panache provides the `PanacheRepository` interface, which should be implemented by all of your repository classes and provides utility methods for most CRUD operations. For this book, we are *not* going to use the repository pattern, since we are not going to build complex queries and this pattern requires creating extra classes. It might be interesting to consider this pattern when your project evolves if your domain needs more complex data access logic.

The active record pattern was first coined by Martin Fowler, and it's defined as follows:

> *An object that wraps a row in a database table or view, encapsulates the database access, and adds domain logic on that data. (Fowler, 2002, p. 160)*

This pattern is most suitable when the domain logic is not too complex, and only simple CRUD operations are performed on the entities. Our application won't need complex queries or data access logic. We'll be using this pattern to implement the persistence layer of our task manager application.

In order to take advantage of the active record pattern and have access to implementations for the most common operations, Panache provides the `PanacheEntity` class, which should be extended by all of the entities. Let's now implement the different entities for our project.

User

The `User` entity will be used for the application's logic, but also the **authentication** and **authorization** services. Let's start by creating a new package that'll contain all the classes related to the user management features.

From the **File** menu, we click on the **New** submenu, and then the **Package** entry.

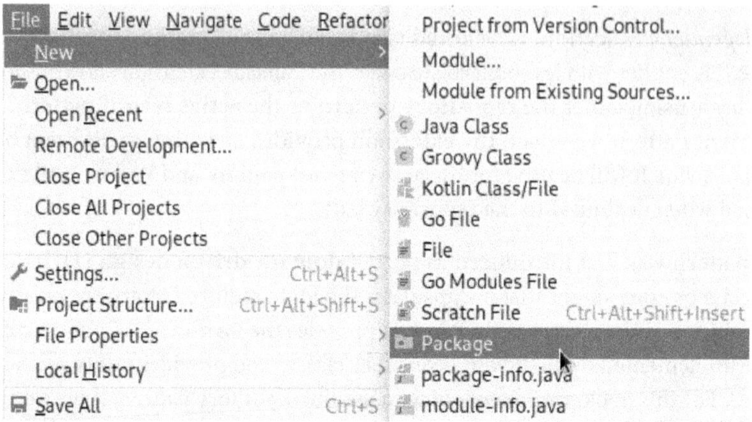

Figure 2.4 – A screenshot of the IntelliJ File | New | Package menu entry

In the modal dialog form, we will type the name of the new package: `com.example.fullstack.user`.

New Package

com.example.fullstack.user

Figure 2.5 – A screenshot of the IntelliJ New Package dialog

As I mentioned in *Chapter 1, Bootstrapping the Project*, you can use a base package name other than `com.example.fullstack`. However, I strongly recommend grouping the different features in individual feature packages (`user`, `project`, `task`, and so on). If the project evolves in the future and needs to be broken down into separate modules or applications, the refactoring process will be much easier.

We can now create the `User` class by right-clicking on the package we just created and then clicking on the **New** submenu and the **Java Class** menu entry.

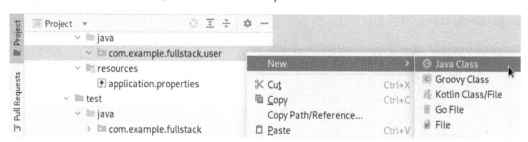

Figure 2.6 – A screenshot of the IntelliJ New | Java Class menu entry

We can now type the name of the Java class that will represent our User entity.

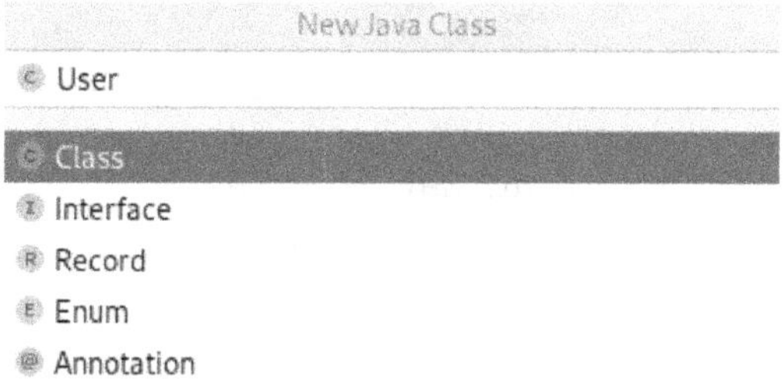

Figure 2.7 – A screenshot of the IntelliJ New Java Class dialog

We can now implement the JPA entity for User; let us analyze some of the code snippets from this class (you can find the complete code in the GitHub repository at https://github.com/ PacktPublishing/Full-Stack-Quarkus-and-React/blob/main/chapter-02/ src/main/java/com/example/fullstack/user/User.java):

```
@Entity
@Table(name = "users")
public class User extends PanacheEntity {
  // omitted field declarations


}
```

The previous code snippet contains the class declaration. The main requirement when declaring JPA entities is to annotate the class with the @Entity annotation. Hibernate will detect this annotation and map it to a database table. We will also customize the name of the table to where the User entity will be mapped; we use the @Table annotation for this purpose, in this case, mapping it to the users table.

For the sake of simplicity, we are taking advantage of the active record pattern provided by Panache. To achieve this, we need to extend the PanacheEntity class. This will provide the entity class with utility methods to perform basic CRUD and query operations. In addition, this allows us to omit the otherwise necessary boilerplate getter and setter methods. We declare the JPA columns as public fields and Panache will add the getter/setter methods at runtime and replace all the field calls with the proper setter or getter method calls. This is another good example of Quarkus' goal to bring back developer joy, removing the need to write redundant and repetitive code. Let us now analyze the class instance variables:

```
@Column(unique = true, nullable = false)
public String name;
```

```
@Column(nullable = false)
String password;

@CreationTimestamp
@Column(updatable = false, nullable = false)
public ZonedDateTime created;

@Version
public int version;

@ElementCollection(fetch = FetchType.EAGER)
@CollectionTable(name = "user_roles", joinColumns =
  @JoinColumn(name = "id"))
@Column(name = "role")
public List<String> roles;
```

You might have noticed that the User class code doesn't contain an @Id annotated JPA column. The PanacheEntity base class provides an id field with an automatically generated long value that will be mapped to a database sequence.

Let's now examine each of the fields we declared for the User class and what their purpose is:

- `public String name;`

  This field represents the username. We want this value to be unique for the application since it will be used as a friendly way to identify the users. In this case, we add the JPA @Column annotation and declare it unique and not nullable.

- `String password;`

  This field will be used to store the user's password and will be used by the application's authentication services. Note that we will encrypt this password using the bcrypt hashing function before storing it in the database; you should *never* store clear text passwords.

  In contrast to the rest of the fields, this one is not public, but package-private. This implies that when User instance objects get serialized through our HTTP API, this field won't be included.

- `public ZonedDateTime created;`

  This field is used for auditing purposes. It's annotated with Hibernate's @CreationTimestamp, which will set the field value to the current Java virtual machine date when saving the entity for the first time. Since this is a required field and should only be set upon creation, we add the @Column annotation with the updatable and nullable options set to false.

- `public int version;`

  This field is used for **optimistic locking**. It will prevent saving an outdated version of the entity that would otherwise overwrite the most recent changes of a newer version. The optimistic locking features are provided by Hibernate and JPA's `@Version` annotation. This field is automatically set by Hibernate upon insertion. It will be checked and automatically incremented upon saving.

- `public List<String> roles;`

  This field will be used by the application's authorization services. Each user can have multiple roles, depending on the operation it's allowed to perform. For the purposes of the task manager application, we'll be defining an `admin` and a `user` role. Users with the `user` role will be able to use the task manager, and users with the `admin` role will be able to manage the application's users.

  This field has three annotations mostly related to where this data is mapped in the database. The `@ElementCollection` annotation is used when mapping very simple non-entity data that is embedded into an entity. This is a very good example: we are simply embedding a collection of strings representing the roles the user has. We've also configured the `EAGER fetch` option since we want to always retrieve the list of roles for the user.

  The `@CollectionTable` annotation is used to configure the target table of embedded types such as this one. In this case, we're specifying the name of the target table, `user_roles`, and the column used to reference the foreign key as a nested `@JoinColumn` annotation.

  The `@Column` annotation is used to configure the name of the column containing the effective roles. However, in this case, it's configuring the name of the column in the target table, `user_roles`.

Now that we've gone over the `User` entity, let us define the `Project` class.

Project

The `Project` entity is used by the task manager users to optionally group tasks. For example, users might be able to create a *Work* project to group the tasks related to their job and a *Home* project to group those related to their daily chores.

Just like we did for users, we'll start by creating a specific package to manage the project. You can follow the same steps we did to create the user package, but we will now create the `com.example.fullstack.project` package instead. We will also create a new Java class in this package with the name `Project`.

Let's implement the code for the `Project` entity. The following snippet contains the most relevant parts; you can find the complete code in the GitHub repository at https://github.com/PacktPublishing/Full-Stack-Development-with-Quarkus-and-React/tree/main/chapter-02/src/main/java/com/example/fullstack/project/Project.java:

```java
@Entity
@Table(
  name = "projects",
  uniqueConstraints = {
    @UniqueConstraint(columnNames = {"name", "user_id"})
  }
)
public class Project extends PanacheEntity {

  @Column(nullable = false)
  public String name;

  @ManyToOne(optional = false)
  public User user;

  @CreationTimestamp
  @Column(updatable = false, nullable = false)
  public ZonedDateTime created;

  @Version
  public int version;
}
```

Just as we did for the `User` class, we annotate the new `Project` class with the `@Entity` annotation and extend the `PanacheEntity` class. We're also going to annotate the class with the `@Table` annotation to specify the table name: `projects`. In addition, we're going to define a unique constraint by configuring the `uniqueConstraints` option of the `@Table` annotation. We want to prevent users from defining duplicate projects, so in this case, we add a unique constraint for the `user_id` and name columns.

Let's now see in detail each of the fields of the `Project` entity:

- `public String name;`

  This field is used to define the name of the project. It's a required field, so we configure it as non-nullable with the `@Column` annotation. We want this field to be unique for *each* user. In this case, instead of using the unique option of the `@Column` annotation, we use a unique constraint. If we were to define the column as unique, then the name would be unique for *all* users, so two users wouldn't be able to each have a project named *Work*.

- `public User user;`

  The user field determines the user who owns this project. It's defined as a reference to the `User` entity through an `@ManyToOne` JPA annotation. Since it's a required field, the `optional` configuration of the annotation is set to `false`, meaning that this field can't be persisted with a null value.

- `public ZonedDateTime created;`

  Just like with the `User` entity, this field is used for auditing purposes to record the timestamp when an instance was created. The same configurations as with the `User` entity apply.

- `public int version;`

  Field used for optimistic locking; we will apply the same configuration as we did for the `User` entity.

Now that we've seen the `User` and `Project` classes, let's analyze `Task`, the remaining JPA entity that we'll need for our data model.

Task

The `Task` class is the main entity of the task manager application data model. Users should be able to define new tasks and track their completion status through the application's interface.

Just like we did for projects and users, we should start by creating a new package called `com.example.fullstack.task` following the procedure we've already seen. We can now create a new Java class called `Task` in this package.

Let us analyze some of the code snippets from the `Task` entity (you can find the complete code in the GitHub repository at `https://github.com/PacktPublishing/Full-Stack-Development-with-Quarkus-and-React/tree/main/chapter-02/src/main/java/com/example/fullstack/task/Task.java`):

```
@Entity
@Table(name = "tasks")
public class Task extends PanacheEntity {
```

```
    // omitted field declarations


}
```

The previous code snippet contains the class declaration. Like for the rest of the application's entities, we'll start by annotating the `Task` class with the `@Entity` annotation and extending the `PanacheEntity` class. Using the `@Table` annotation, we'll configure the name of the target database table, `tasks`. The next code snippet contains the class instance variables:

```
@Column(nullable = false) public String title;


@Column(length = 1000) public String description;


public Integer priority;
@ManyToOne(optional = false) public User user;
public ZonedDateTime complete;


@ManyToOne public Project project;


@CreationTimestamp
@Column(updatable = false, nullable = false)
public ZonedDateTime created;


@Version public int version;
```

Let's now examine the fields for this class:

- `public String title;`

  This field is used to define the task title. It's employed to provide a short description of the task that will be presented to the users in summary tables. This is a required field, so it's annotated with the `@Column` annotation and the `nullable` option set to `false`.

- `public String description;`

  The description is utilized by users to provide a more detailed definition of the task. Hibernate maps strings to `varchar(255)` database data types. Since this length might be insufficient, we'll manually set the column length to `1000` by using the `@Column` annotation with the `length` option.

- `public Integer priority;`

  This field allows users to optionally sort and prioritize tasks. The lower the value, the higher the priority the task will have. Since this field is optional and we don't want to provide further customizations, it doesn't require any annotation or configuration.

- `public User user;`

  The user field determines the user who owns this task. The mapping and configuration are the same as for the `Project` entity.

- `public ZonedDateTime complete;`

  Once users consider a task as complete, they can use the task manager application to set this field with a timestamp. If the field is null, the task is considered to be incomplete.

- `public Project project;`

  This is an optional field; users can use it to group tasks by project. Similar to the user field, we use the @ManyToOne annotation to define the relationship with the `Project` entity.

Same as `Project` and `User`, this class also has `created` and `version` audit fields, which we've already seen how to configure.

Now that we've implemented all of the entities for the task manager, let's see how Hibernate maps them into database tables.

Database tables resulting from the ORM entities

In the following diagram, you can see the resulting tables and relationships that will be created in the database when the application starts:

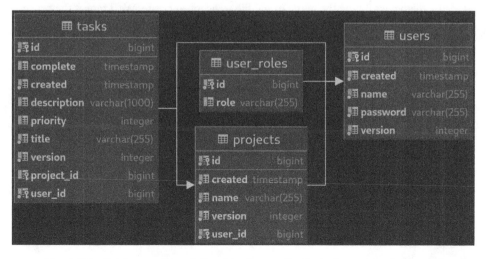

Figure 2.8 – Database entity-relationship diagram for the task manager data model

We only defined three entities; however, there are four tables in total. There is a table per entity and an extra table, `user_roles`, which is used to store the `User` entity embedded values for the `roles` collection field.

Loading the application's initial data

In the *Configuring Quarkus* section, we added two entries to the `application.properties` file to configure the files containing SQL scripts that will be loaded upon application startup. We used the `quarkus.hibernate-orm.sql-load-script` configuration option for this purpose and configured the `import.sql` file for the production environment and `import-dev.sql` for the development environment. Let's now create these files and analyze their content.

To create the files, we can right-click on the **resources** folder, click on the **New** submenu, and then select the **File** menu entry.

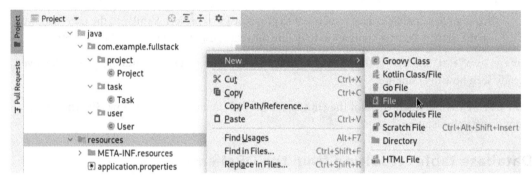

Figure 2.9 – A screenshot of the IntelliJ New | File menu entry

In the **New File** dialog, we'll type the name of the file we want to create. We'll repeat the procedure twice, once for the `import.sql` file and another time for the `import-dev.sql` file.

Figure 2.10 – A screenshot of the IntelliJ New File dialog

The `import.sql` file contains SQL statements to create a user named `admin` with the `admin` and `user` roles. You can check the content of this file in the book's GitHub repository.

The `import-dev.sql` file has the same content as the `import.sql` file, additional entries to define an extra user named `user` with the `user` role, and an initial project for this user with the name `Work`. This initial data will be handy when we start implementing the frontend of the application since it will populate the different screens with sample data. Feel free to add more fake data to `import-dev.sql` if it helps you during the next development iterations.

We've now implemented all of the necessary entities and configurations for the persistence layer of the task manager application. Let's now dig into Quarkus Dev Services and how we can start our application without the need to configure or deploy a database instance.

Quarkus Dev Services

Quarkus is deeply focused on improving the developer experience for Java – we've seen multiple examples so far. Quarkus Dev Services is a step further in this quest for developer joy. Its main feature is to automatically provision services in development and test modes. This means that if your project has an extension configured to provide a database service, a messaging provider, an in-memory datastore, or one of the many other supported services, Quarkus will automatically start and configure this service for your application upon its startup.

Under the hood, Quarkus uses Testcontainers and Docker to provide these services. It's required to have a Docker-compatible environment for this feature to work.

This means we won't need to have a local PostgreSQL database available when running our application in development mode. We won't need to provide a configuration either. Quarkus will start a Docker container with a fresh PostgreSQL database and configure the application to use it automatically. When we start the application with the `./mvnw quarkus:dev` command, we should see the following:

Figure 2.11 – A screenshot of quarkus:dev with the started postgresql Dev Services

Quarkus Dev Services is also very convenient for testing purposes since the tests will run in a real database with a clean slate. This means you won't need to configure mocks or in-memory databases or provide a specific testing database environment to run your tests. Just like with dev mode, when you run your tests, Quarkus will start the development services for you without any further configuration.

Of course, you can opt-out of this behavior and configure a more traditional workflow with a fixed development environment. In that case, you just need to add the following entry to the `application.properties` file: `quarkus.devservices.enabled=false`. However, simply providing a manual service configuration should also disable Dev Services.

Summary

In this chapter, we've seen how to add a fully reactive persistence layer to our application using Hibernate Reactive with Panache and the Reactive PostgreSQL client. We started by adding the required dependencies to our project using the Quarkus Maven plugin and then providing the required configuration for both the production and development environments. Then, we implemented the entities that we'll be using in our task manager application and went over the different JPA and Hibernate Java annotations we used to configure them. We also examined Quarkus Dev Services and what advantages it provides both for the development mode and when running tests.

You should now be able to provide a basic persistence layer for your applications. In the next chapter, we'll create the HTTP API that will be consumed by the frontend part of the application. We'll create the different services that'll consume the data layer we just developed and implement the required HTTP endpoints.

Questions

1. What is Hibernate and how does it relate to JPA?

2. Is it possible to implement a fully reactive persistence layer with Quarkus?

3. How do you implement an entity with the active record pattern in Quarkus?

4. How do you load initial application data to the database?

5. What is Quarkus Dev Services?

3
Creating the HTTP API

In this chapter, we will create the HTTP API that will be consumed by the frontend of our task manager. We'll start by learning about how to write HTTP REST endpoints in Quarkus. Then, we'll define the services that will encapsulate the business logic of our application and make use of the data layer we implemented in *Chapter 2, Adding Persistence*. Next, we will implement the endpoints that will expose the functionality of the different services to the frontend application.

By the end of this chapter, you should be able to implement HTTP and **representational state transfer** (**REST**) endpoints in Quarkus. You should also be able to create singleton services and use dependency injection to provide their instances to the endpoint implementation classes.

We will be covering the following topics in this chapter:

- Writing HTTP REST endpoints in Quarkus
- Implementing the task manager business logic
- Exposing the task manager to the frontend
- Dealing with service exceptions

Technical requirements

You will need the latest Java JDK LTS version (at the time of writing, Java 17). In this book, we will be using Fedora Linux, but you can use Windows or macOS as well.

You will need a working Docker environment to take advantage of *Quarkus Dev Services*. There are Docker packages available for most Linux distributions. If you are on a Windows or macOS machine, you can install Docker Desktop.

If you're not using IntelliJ IDEA Ultimate, you'll need a tool such as **cURL** or **Postman** to interact with the HTTP endpoints implemented.

You can download the full source code for this chapter from `https://github.com/PacktPublishing/Full-Stack-Quarkus-and-React/tree/main/chapter-03`.

Writing HTTP REST endpoints in Quarkus

Quarkus provides several ways to implement HTTP and REST endpoints. In *Chapter 1*, *Bootstrapping the Project*, we learned about the imperative and reactive paradigms and how Quarkus can be used for both approaches. In this book, we are following the reactive approach to take advantage of its improved performance.

In the *Bootstrapping a Quarkus application* section of *Chapter 1*, *Bootstrapping the Project*, we initialized the project and added *RESTEasy Reactive* to the list of dependencies. This dependency is what will allow us to implement the reactive HTTP endpoints. As we learned, RESTEasy provides an implementation of *JAX-RS* based on *Vert.x*. One of the major advantages of RESTEasy Reactive compared to the regular RESTEasy alternatives is that it allows us to implement both blocking and non-blocking endpoints.

Although we already have the RESTEasy dependency, we'll need additional dependencies to be able to serialize our database entities into **JSON** so that they can be consumed by the HTTP API clients.

Adding the required dependencies to our project

To add the missing dependencies, we can use the `quarkus:add-extension` Maven goal, which will automatically insert the required code in the project's `pom.xml` file. Running the following command in the project's root will add the dependencies for you:

```
./mvnw quarkus:add-extension -Dextensions=resteasy-reactive-jackson
```

Once executed, you should see the following message:

```
[INFO] --- quarkus-maven-plugin:2.7.2.Final:add-extension (default-cli) @ reactive ---
[INFO] [SUCCESS] ✔ Extension io.quarkus:quarkus-resteasy-reactive-jackson has been installed
```

Figure 3.1 – A screenshot of the execution result of the quarkus:add-extension command

In this case, we've only added a single dependency, `quarkus-resteasy-reactive-jackson`, which provides serialization support for RESTEasy Reactive using the **Jackson** library. Our `pom.xml` file should now contain an additional entry in the dependencies section:

```
<dependency>
  <groupId>io.quarkus</groupId>
```

```
  <artifactId>quarkus-resteasy-reactive-jackson
  </artifactId>
</dependency>
```

Now that the project is ready, let us see how we can implement both blocking and non-blocking endpoints with RESTEasy Reactive.

> **Jackson**
>
> Jackson is one of the most popular Java libraries, commonly known for its JSON data format serialization and deserialization features. However, Jackson is much more than that and offers a complete suite of data processing tools with support for many other data formats and encodings.

Writing a blocking synchronous endpoint

To implement a blocking endpoint, you just need to follow the regular JAX-RS convention and use the annotations provided. You can write the following code snippet to define a blocking endpoint in Quarkus:

```
@Path("blocking-endpoint")
public class BlockingEndpoint {
  @GET
  public String hello() {
    return "Hello World";
  }
}
```

We start by defining the HTTP path using the `@Path` annotation – in this case, `blocking-endpoint`. This annotation will also allow Quarkus to discover the class and expose its JAX-RS-annotated methods as HTTP endpoints.

The next annotation, `@GET`, is used to indicate that the annotated method should be used to respond to HTTP GET requests. This annotation is used on a `hello()` method that returns a `String` instance. Note that the `return` type of the method is what will dictate whether Quarkus treats this as a blocking or a non-blocking endpoint. In this case, once the application is started, an HTTP request to `localhost:8080/blocking-endpoint` will return a `text/plain` response with a simple body, `Hello World`.

If you are already familiar with JAX-RS, there's nothing new here – you should already be familiar with this code. Now, let us see how to implement the same endpoint in a non-blocking, more performant way.

Writing a non-blocking asynchronous endpoint

The implementation of the non-blocking version of the endpoint just requires a minor change to the return type of the method signature. You can use the following code snippet to define a non-blocking endpoint in Quarkus:

```
@Path("non-blocking-endpoint")
public class NonBlockingEndpoint {
  @GET
  public Uni<String> hello() {
    return Uni.createFrom().item("Hello").onItem().
      transform(s -> s + " World");
  }
}
```

In this case, we've defined the endpoint using a different path: `non-blocking-endpoint`. Regarding the JAX-RS annotations, nothing has changed – both the class and the method keep the annotations we used for the `blocking-endpoint` snippet.

However, the method return type and its implementation have changed. Now, instead of returning a regular `String` instance, the method returns a `Uni<String>` object. **Uni** is part of **Mutiny**, an event-driven reactive programming library for Java. Mutiny is integrated into Quarkus and is the primary model used when dealing with reactive types.

> **Mutiny types**
>
> In Mutiny, `Uni` represents a lazy asynchronous action that emits a *single* event. This event can either be an item or a failure. Mutiny provides two types, the `Uni` type and the **Multi** type. `Multi` is an asynchronous action, just like `Uni`, but emits *multiple* events instead. Both types are used as the starting point and input for a Mutiny pipeline. Users can then perform asynchronous operations on the items emitted by the pipeline in the processing part and, finally, subscribe to them.

In the code snippet, we use the following code to create a `Uni` object: `Uni.createFrom().item("Hello")`. This is something that you wouldn't generally do since we are just creating an asynchronous pipeline for an item that is already available. When we dive into the endpoint implementations, we'll see how these events can be consumed directly from the asynchronous events that a database emits when you perform a query. In the snippet, we're also applying a transformation to append `" World"` to the item provided. The method returns the resulting transformed `Uni`; Quarkus will then take care of subscribing to the pipeline and convert it to an applicable HTTP response. Just as in the blocking example, an HTTP request to `localhost:8080/non-blocking-endpoint` will return a `text/plain` response with a simple body, `Hello World`.

Now that we know how to implement non-blocking endpoints in Quarkus, taking advantage of the reactive paradigm, let us see how to implement the business logic that we'll later expose as HTTP endpoints.

Implementing the task manager business logic

In *Chapter 2*, *Adding Persistence*, we created the persistence layer for the application. We created entities for users, projects, and tasks following the active record pattern. In this chapter, we're creating the HTTP API that will be consumed by the frontend of our task manager application. However, before implementing the HTTP endpoints, it's a good practice to encapsulate the business logic of the application within different service classes. We can expose the operations provided by these services later by implementing the non-blocking JAX-RS annotated classes and methods.

We are going to implement three services: UserService, TaskService, and ProjectService. Let us start by analyzing UserService since it contains some methods that will be reused by the rest of the services.

UserService

The user service will be used to encapsulate all the required business logic for managing the application's users. Later on, when we implement the task manager's security, we will also use this service to retrieve the currently logged-in user.

We'll start by creating a new class, UserService, by right-clicking on the com.example.fullstack.user package and clicking on the **New** submenu and the **Java Class** menu entry:

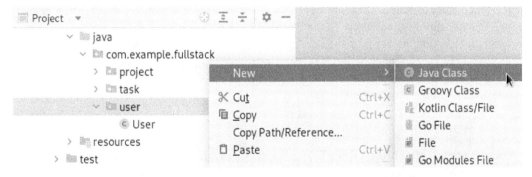

Figure 3.2 – A screenshot of the Java Class menu entry in IntelliJ

We can now type the name of the new Java class where we'll implement the user service business logic.

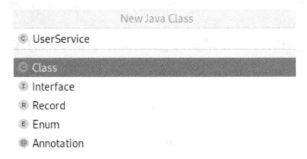

Figure 3.3 – A screenshot of the New Java Class dialog in IntelliJ

In the following code snippet, you will find the relevant source code for the class declaration (you can find the complete code in the GitHub repository at `https://github.com/PacktPublishing/Full-Stack-Quarkus-and-React/blob/main/chapter-03/src/main/java/com/example/fullstack/user/UserService.java`):

```
@ApplicationScoped
public class UserService {
  // ...
}
```

In *Chapter 1*, *Bootstrapping the Project*, we learned that Quarkus uses *ArC*, a *CDI* implementation to provide dependency injection. There are several ways to declare **beans** in Quarkus, the easiest being the use of bean-defining annotations. In the `UserService` code snippet, you'll notice that we've annotated the class with the `@ApplicationScoped` annotation. This is a bean-defining annotation that declares a singleton instance of the Java class, which will be shared in a single application context. Whenever `UserService` gets injected into other beans, the CDI container will inject the same instance to each of these dependent beans.

> **Bean**
>
> A bean is just an object managed by a CDI container that supports a set of services such as life cycle callbacks, dependency injection, and interceptors. The CDI container is the environment where the application runs. It's responsible for the creation and destruction of bean instances and the injection of these instances into other beans. You can learn more about the different CDI scopes in the official Quarkus guide here: `https://quarkus.io/guides/cdi`.

Now, let us continue by implementing and analyzing the methods that the `UserService` class will provide:

- `findById(long id)`:

```
public Uni<User> findById(long id) {
  return User.<User>findById(id)
    .onItem().ifNull().failWith(() -> new
    ObjectNotFoundException(id, "User"));
}
```

This method will return a `Uni` object that will either emit a `User` entity for the ID provided if the user exists in the database or fail with an exception if it doesn't. The implementation starts by calling the `User.findById` static method, which is part of the active record pattern methods provided by the `PanacheEntity` base class. As we saw in *Chapter 2*, *Adding Persistence*, the `PanacheEntity` class provided many convenient features to improve our development experience. In this case, we're taking advantage of one of the several provided static methods to perform operations on the entity. In addition, we don't need to inject anything into the service class since all these methods can be accessed statically.

The `User.findById` method call returns a `Uni` object that emits an item containing either the found user or `null` if it doesn't exist. However, in this case, we don't want to emit a `null` value. We *asynchronously* process the result of the `findById` method call to check for the `null` value, `onItem().ifNull()`, and throw a Hibernate `ObjectNotFoundException` with the ID of the missing user. If a user subscribes to this `Uni` and nothing is found, the subscription will fail instead of emitting a `null` value.

- `findByName(String name)`:

```
public Uni<User> findByName(String name) {
  return User.find("name", name).firstResult();
}
```

This method returns a `Uni` object that emits a `User` entity for the matching name or `null` if no user matches the criteria. In the method implementation, we're once again taking advantage of one of the static query methods provided by `PanacheEntity`: `User.find("name", name)`. The first argument is the **Hibernate Query Language** (**HQL**) query and the second is the parameter to be passed to the query. However, in this case, we're not even providing a complete HQL query but one of the simplified forms that Quarkus supports. This query is the equivalent of the HQL: `from User where name = ?`.

You'll also notice that we're returning the first available result. The `firstResult` method call produces a `Uni` instance that either emits the first `User` entity matching the criteria or `null`. We can be sure there will either be a single result or none since we declared the `name` field as unique in its `@Column` definition within the `User` class.

Unlike with the previous findById method, we are not checking for a null item to fail with an exception. This method will be consumed by the authorization service that we'll implement in *Chapter 4*, *Securing the Application*; the null value processing will be done there instead.

- list():

```
public Uni<List<User>> list() {
  return User.listAll();
}
```

This will return a Uni object that emits a list containing all of the available entities in the database. The method implementation returns the result of the delegated call to the User. listAll() method provided by PanacheEntity.

- create(User user):

```
@ReactiveTransactional
public Uni<User> create(User user) {
  user.password = BcryptUtil.bcryptHash
    (user.password);
  return user.persistAndFlush();
}
```

This method persists a new user entity in the database with the data from the provided User object instance and returns a Uni object with a single item containing the updated User object. Before persisting the object, we update the password value in the provided object with a *bcrypt* hash. For this purpose, we will use the BcryptUtil.bcryptHash method using the originally provided plaintext password as input.

Once the hashed password is set, we are ready to persist the new instance by invoking the persistAndFlush() method. This is another of the methods provided by the PanacheEntity base class.

The method is annotated with a @ReactiveTransactional annotation that indicates that the executed method should be run within a reactive Mutiny transaction to persist in the database. We will configure all of the service methods that perform write or delete operations with this annotation to make sure that the data is only persisted in the case that the complete service logic covered by the transaction is successful.

---

bcrypt

bcrypt is a password-hashing function based on the Blowfish cipher. bcrypt is especially recommended for hashing passwords because it is a slow and expensive algorithm. The slowness of the function makes it ideal to store passwords because it helps mitigate brute-force attacks by reducing the number of hashes per second an attacker can use when performing a dictionary attack.

- update(User user):

```
@ReactiveTransactional
public Uni<User> update(User user) {
  return findById(user.id)
    .chain(u -> User.getSession())
    .chain(s -> s.merge(user));
}
```

This method will update the data of the user entity in the database with the values from the provided user entity input argument. Since this is a write operation, the method is also annotated with the @ReactiveTransactional annotation we already covered.

The supplied user argument will most likely come from a deserialized user entity from an HTTP request body. So, in this case, since the entity will be in a detached state and we want to update it, we can't call the user.persistAndFlush() method we used for the create method.

The method implementation starts by calling the findById method we implemented previously, so if the user doesn't exist, the Uni instance will emit a failure.

The method implementation continues by chaining the Uni instance and mapping the emitted item to the underlying Hibernate session, .chain(u -> User.getSession()). The chain method is how Mutiny chains *asynchronous* operations and maps their result. It expects a lambda expression that accepts the resolved item and should return a new Uni instance with the mapped value. In this case, we are ignoring the received user and returning a Uni<Session> object.

Next, we invoke the merge method of the now available Hibernate session. The merge operation will copy the state of the provided detached user instance onto a managed instance of the same entity. Internally, Hibernate will retrieve the user from the database first and then copy all the attribute values of the detached version. In addition, Hibernate will check the @Version field to enforce optimistic locking and prevent a more recent version of the entity from being overwritten. The method returns the updated, managed entity and makes it available to the Uni subscriber.

Note that right now, this method will allow the user's password to be updated from the API consumer. In *Chapter 4, Securing the Application*, we'll add another service method to provide the user password update functionalities and modify this one to ignore the provided password field if present.

- delete(long id):

```
@ReactiveTransactional
public Uni<Void> delete(long id) {
  return findById(id)
```

```
        .chain(u -> Uni.combine().all().unis(
            Task.delete("user.id", u.id),
            Project.delete("user.id", u.id)
        ).asTuple()
        .chain(t -> u.delete())
    );
}
```

This method deletes the user with the `id` provided by the database and all of its owned tasks
and projects and returns a `Uni` object that will emit a `null` item if it completes successfully.
Just as with the `update` method, the implementation starts by calling the `findById` method
we implemented previously so that we can reuse the "not found" logic. The user deletion should
also be covered by a transaction since it will perform a persistent delete operation in the database
– we must annotate the method with `@ReactiveTransactional` too.

The `Uni` instance returning the found user is then chained to a more complex lambda expression
that performs asynchronous deletions of the associated tasks and projects. To delete all tasks
and projects, we're performing two different batch deletions: `Task.delete("user.id",
u.id)` and `Project.delete("user.id", u.id)`. In both cases, we're issuing a
batch delete operation bound to an HQL query that filters out entities that have a user ID
that matches the one provided. To make sure that the user is only deleted once every related
task and project is deleted, we are combining both delete operations into a single `Uni` object:
`Uni.combine().all().unis(...).asTuple()`. This operation is then chained to
another lambda expression that will be responsible for the effective user deletion if the previous
operation succeeds.

- `getCurrentUser()`:

```
public Uni<User> getCurrentUser() {
    // TODO: replace implementation once security is
        added to the project
    return User.find("order by ID").firstResult();
}
```

This is an initial dummy implementation of the `getCurrentUser` method that is needed
by other services. Until we implement the application security, this method just returns a `Uni`
instance that emits a single item containing the user with the lowest ID available in the database.

Now that we've seen the implementation details of `UserService`, let us see the implementation
of `ProjectService`.

ProjectService

The project service encapsulates all of the business logic related to the management of projects for our task manager application. Just as we did for users, we'll start by creating a new `ProjectService` class in the `com.example.fullstack.project` package. The following code snippet contains the relevant parts of the source code for the class declaration and its constructor (you can find the complete code in the GitHub repository at `https://github.com/PacktPublishing/Full-Stack-Development-with-Quarkus-and-React/tree/main/chapter-03/src/main/java/com/example/fullstack/project/ProjectService.java`):

```
@ApplicationScoped
public class ProjectService {

  private final UserService userService;

  @Inject
  public ProjectService(UserService userService) {
    this.userService = userService;
  }
  // …
}
```

Just as with `UserService`, this class is annotated with an `@ApplicationScoped` annotation so that it can be injected as a singleton into other beans. When dealing with projects in our application, we'll need to know the currently logged-in user to perform operations that only apply to this user. `UserService` is responsible for providing this information, so we need to inject it into this service. In this case, we are using constructor-based injection. The constructor is annotated with `@Inject` and has a single parameter, `UserService userService`. When Quarkus instantiates this class, the singleton `UserService` instance from the application context will automatically be injected.

Now, let us analyze the different method implementations, focusing on those that differ from the analogous methods we implemented in the `UserService` class:

- `findById(long id)`:

```
public Uni<Project> findById(long id) {
  return userService.getCurrentUser()
    .chain(user -> Project.<Project>findById(id)
      .onItem().ifNull().failWith(() -> new
        ObjectNotFoundException(id, "Project"))
```

```
        .onItem().invoke(project -> {
        if (!user.equals(project.user)) {
            throw new UnauthorizedException("You are not
            allowed to update this project");
        }
    }));
}
```

This method returns a `Uni` instance that will emit an item containing the project with the requested ID. If the project doesn't exist, the `Uni` subscription will fail with `ObjectNotFoundException`. If the project belongs to a user other than the one logged in, it will fail with `UnauthorizedException`.

We start the implementation by calling the `getCurrentUser` method exposed by `UserService`. The returned `Uni` instance that will emit the logged-in user is then chained to a lambda expression, which receives this user as its only argument and retrieves the project by its ID. The lambda expression contains asynchronous checks for the retrieved project. Just as we did for users, we check for a `null` item to throw an exception. In addition, this time, we also check that the project belongs to the returned user and we throw `UnauthorizedException` if it doesn't.

- `listForUser()`:

```
public Uni<List<Project>> listForUser() {
    return userService.getCurrentUser()
        .chain(user -> Project.find("user", user).list());
}
```

This method returns a `Uni` object that will emit an item containing a list of all the project entities in the database that belong to the currently logged-in user.

The implementation starts by invoking `getCurrentUser` to retrieve a `Uni` object that will emit the user performing this operation. The `Uni` instance is *asynchronously* chained to a lambda expression that uses the input user argument as the parameter of an HQL query to find all users.

- `create(Project project)`:

```
@ReactiveTransactional
public Uni<Project> create(Project project) {
    return userService.getCurrentUser()
        .chain(user -> {
            project.user = user;
            return project.persistAndFlush();
```

```
    });
  }
```

The method creates a new project in the database with the data provided in the input project argument and returns a `Uni` object that will emit an item with the new project.

We start the implementation by retrieving the currently logged-in user. Then we perform an asynchronous transformation to override the provided project's user with the one received from the `Uni` subscription and store it in the database using the `persistAndFlush` method. With this, we make sure that the project is assigned to the currently logged-in user and not to whatever value was provided as input.

- `update(Project project)`:

```
@ReactiveTransactional
public Uni<Project> update(Project project) {
  return findById(project.id)
    .chain(p -> Project.getSession())
    .chain(s -> s.merge(project));
}
```

This method will update the data of the project entity in the database with the values from the provided `project` entity input argument. The implementation follows the same pattern described for the analogous operation in `UserService`.

- `delete(long id)`:

```
@ReactiveTransactional
public Uni<Void> delete(long id) {
  return findById(id)
    .chain(p -> Task.update("project = null where
      project = ?1", p)
      .chain(i -> p.delete()));
}
```

This method deletes the project with the matching ID from the database and updates all of the related tasks to unset the project. The method returns a `Uni` object that will emit a `null` item if it completes successfully. The implementation follows a similar approach to the one described for the `delete` method in `UserService`. However, in this case, instead of performing a batch delete of tasks, we perform a batch update using a simplified HQL query: `Task.update("project = null where project = ?1", p)`.

We've already covered the `UserService` and `ProjectService` implementations – now, let us analyze the code for `TaskService`.

TaskService

The task service contains the complete business logic to manage the task entities in our task manager. Just as we did for users, we'll start by creating a new `TaskService` class in the `com.example.fullstack.task` package. The following code snippet contains the relevant parts of the source code for the class declaration and its constructor (you can find the complete code in the GitHub repository at `https://github.com/PacktPublishing/Full-Stack-Development-with-Quarkus-and-React/tree/main/chapter-03/src/main/java/com/example/fullstack/task/TaskService.java`):

```
@ApplicationScoped
public class TaskService {

  private final UserService userService;

  @Inject
  public TaskService(UserService userService) {
    this.userService = userService;
  }
  // ...
}
```

The `TaskService` implementation is very similar to the one we did for `ProjectService`. It's a singleton `@ApplicationScoped` bean with `findById`, `listForUser`, `update`, and `delete` methods too.

Now, let's check the methods of the classes and analyze those that are implemented differently from the analogous versions in `ProjectService` and `UserService`:

- `findById(long id)`:

```
public Uni<Task> findById(long id) {
  return userService.getCurrentUser()
    .chain(user -> Task.<Task>findById(id)
      .onItem().ifNull().failWith(() -> new
        ObjectNotFoundException(id, "Task"))
      .onItem().invoke(task -> {
        if (!user.equals(task.user)) {
          throw new UnauthorizedException("You are not
            allowed to update this task");
        }
```

```
    }));
}
```

This method returns a `Uni` object that will emit an item containing a task with the requested ID or will fail with one of the following two exceptions: `ObjectNotFoundException` if the task doesn't exist, or `UnauthorizedException` if the task belongs to a user other than the one logged in.

- `listForUser()`:

```
public Uni<List<Task>> listForUser() {
    return userService.getCurrentUser()
        .chain(user -> Task.find("user", user).list());
}
```

This method returns a `Uni` object that emits an item containing a list of all the task entities available in the database that belong to the currently logged-in user.

- `create(Task task)`:

```
@ReactiveTransactional
public Uni<Task> create(Task task) {
    return userService.getCurrentUser()
        .chain(user -> {
            task.user = user;
            return task.persistAndFlush();
        });
}
```

This method creates a new task in the database with the data provided in the input `task` argument and returns a `Uni` object that will emit an item containing the newly created `task` entity.

- `update(Task task)`:

```
@ReactiveTransactional
public Uni<Task> update(Task task) {
    return findById(task.id)
        .chain(t -> Task.getSession())
        .chain(s -> s.merge(task));
}
```

This method will update the data of the task entity in the database with the values from the `task` entity input argument provided.

- `delete(long id)`:

```
@ReactiveTransactional
public Uni<Void> delete(long id) {
  return findById(id)
    .chain(Task::delete);
}
```

This method deletes the task with the matching ID from the database and returns a `Uni` object that will emit a `null` item if it completes successfully. Unlike for `UserService` and `ProjectService`, tasks have no dependent entities, so we perform the deletion of the entity by chaining it after the `findById` method call.

- `setComplete(long id, boolean complete)`:

```
@ReactiveTransactional
public Uni<Boolean> setComplete(long id, boolean
complete) {
  return findById(id)
    .chain(task -> {
      task.complete = complete ? ZonedDateTime.now() :
        null;
      return task.persistAndFlush();
    })
    .chain(task -> Uni.createFrom().item(complete));
}
```

This method updates the task entity with the matching ID from the database by setting its `complete` field value. If the provided Boolean `complete` argument is `true`, then we set the task's `complete` field with the current timestamp; if it is `false`, we set it to `null`.

Users can achieve a similar result by invoking the `update` method that is also provided. However, we provide this convenient method to encapsulate the `complete` field's business logic and set its value with the current system's time. In addition, this method only focuses on the `complete` field, disregarding optimistic locking.

Now that we've implemented the complete business logic for the task manager application, let us see how to expose it to the frontend by implementing the HTTP endpoints.

Exposing the task manager to the frontend

We have implemented the required services for the task manager. Now, we can use the techniques we learned in the previous section, *Writing a non-blocking asynchronous endpoint*, to expose them to the frontend. Following the same pattern we used for the service implementation, we are going to create three resource controller classes, one for each entity: UserResource, ProjectResource, and TaskResource.

UserResource

This resource will expose the public operations of the UserService class, allowing the HTTP API consumers to perform actions dealing with the users of the task manager. We'll start by creating a new UserResource class in the com.example.fullstack.user package. The following code snippet contains the relevant part to declare the newly created class and its constructor (you can find the complete code in the GitHub repository at https://github.com/PacktPublishing/Full-Stack-Quarkus-and-React/blob/main/chapter-03/src/main/java/com/example/fullstack/user/UserResource.java):

```
@Path("/api/v1/users")
public class UserResource {

  private final UserService userService;

  @Inject
  public UserResource(UserService userService) {
    this.userService = userService;
  }
  // ...
}
```

Just as we learned about in the *Writing a blocking synchronous endpoint* section, to declare and expose an endpoint in Quarkus, you need to annotate the class with the @Path annotation. We'll expose all of the user-related endpoints with the /api/v1/users prefix. We'll follow the same pattern for the rest of the resources and reuse the /api/v1 prefix. This will allow us to distinguish the API endpoints from other kinds of endpoints and also facilitate the implementation of new versions in the future.

Since most of the endpoints will produce a JSON response body, we could also annotate the class with a @Produces(MediaType.APPLICATION_JSON) annotation. However, this is not necessary for Quarkus because the JSON response media type is inferred from the presence of the quarkus-resteasy-reactive-jackson dependency and is used as the default for most of the return types. You can disable this feature by providing the following property: quarkus.resteasy-json.default-json=false.

All of the operations are delegated to UserService, so we'll use constructor-based injection once again to initialize the userService instance variable. Now, let us analyze the class methods and their annotations:

- get():

```
@GET
public Uni<List<User>> get() {
    return userService.list();
}
```

This method is just annotated with the @GET annotation, which indicates that this method should be used to respond to HTTP GET requests. Its implementation delegates the call to the userService.list() method. Since the method is not annotated with an additional path annotation, when running the application, the endpoint will be available at the /api/v1/users URL. When invoked, the user will receive a JSON list of all the available users.

If you start the application with ./mvnw quarkus:dev, you should be able to invoke the following cURL command:

curl localhost:8080/api/v1/users

If everything goes well, you should be able to see something similar to this:

```
$ curl localhost:8080/api/v1/users
[{"id":0,"name":"admin","created":"2022-04-03T10:21:04.804908+02:00","version":0,"roles":["admin","user"]},
{"id":1,"name":"user","created":"2022-04-03T10:21:04.810769+02:00","version":0,"roles":["user"]}]
```

Figure 3.4 – A screenshot of the result of executing cURL to retrieve all users

Note that since the password field in the User entity is not public, it is *not* exposed through the HTTP API.

- create(User user):

```
@POST
@Consumes(MediaType.APPLICATION_JSON)
@ResponseStatus(201)
public Uni<User> create(User user) {
```

```
        return userService.create(user);
    }
```

This endpoint exposes the functionality of `UserService` to create new users in the database. The method contains several annotations. The first one, `@POST`, is used to indicate that this method should be selected to respond to HTTP POST method calls. The `@Consumes(MediaType.APPLICATION_JSON)` annotation indicates that the HTTP request should include a body with an `application/json` content type. This annotation, in combination with the `User user` method parameter, configures Quarkus to deserialize the provided JSON body in the request into a `User` entity instance. The `@ResponseStatus(201)` annotation forces Quarkus to respond with a `201 Created` response status code instead of the standard `200 OK` upon success.

Let's test the endpoint as it is by invoking the following request:

```
curl -X POST
-d"{\"name\":\"packt\",\"password\":\"pass\"}" -H
"Content-Type: application/json" localhost:8080/api/v1/
users
```

The response should show a failure, indicating that the password is a required field. In the next section, *Deserializing the User entity's password field*, we will fix this problem.

* `get(@PathParam("id") long id)`:

```
@GET
@Path("{id}")
public Uni<User> get(@PathParam("id") long id) {
    return userService.findById(id);
}
```

This endpoint returns a JSON representation of the user with the requested ID. In addition to the `@GET` annotation, we've included a `@Path("{id}")` annotation too. The path includes an `{id}` parameter so that we can retrieve the user ID from the request URL. This annotation is used in combination with the annotated method argument: `@PathParam("id") long id`. We can now test the endpoint by invoking the following command:

```
curl localhost:8080/api/v1/users/0
```

If everything goes well, you should be able to see the following response:

```
$ curl localhost:8080/api/v1/users/0
{"id":0,"name":"admin","created":"2022-04-03T10:21:04.804908+02:00","version":0,"roles":["admin","user"]}
```

Figure 3.5 – A screenshot of the result of executing cURL to retrieve user 0

- update(@PathParam("id") long id, User user):

```
@PUT
@Consumes(MediaType.APPLICATION_JSON)
@Path("{id}")
public Uni<User> update(@PathParam("id") long id, User
  user) {
  user.id = id;
  return userService.update(user);
}
```

This endpoint allows us to update the information for the user that matches the id provided with the information provided in the user entity. It's annotated with the @PUT annotation, so Quarkus will select this method when dealing with HTTP PUT requests targeting this path. We've already covered the @Path, @Consumes, and @PathParam annotations when examining the other methods, so I won't explain them again. Note that in the method implementation, we're overriding whatever ID came in the deserialized User entity with the one provided in the HTTP URL path.

- delete(@PathParam("id") long id):

```
@DELETE
@Path("{id}")
public Uni<Void> delete(@PathParam("id") long id) {
  return userService.delete(id);
}
```

This endpoint can be used to delete the user that matches the id provided in the URL path. It's annotated with the @DELETE annotation, so the method will be selected when dealing with HTTP DELETE requests. We can test the endpoint by executing the following command:

`curl -X DELETE -i localhost:8080/api/v1/users/1`

If everything goes well, you should see the following message containing the HTTP response header – the user should no longer exist in the database:

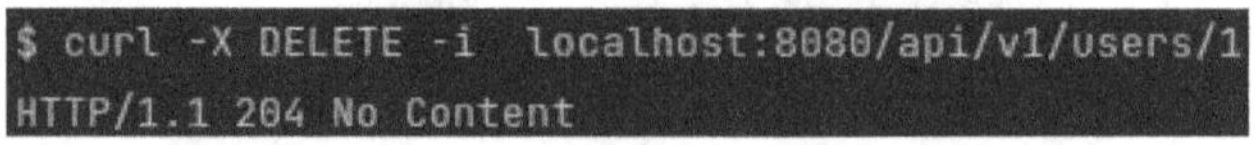

```
$ curl -X DELETE -i localhost:8080/api/v1/users/1
HTTP/1.1 204 No Content
```

Figure 3.6 – A screenshot of the result of executing cURL to delete user 1

- getCurrentUser():

```
@GET
@Path("self")
```

```
public Uni<User> getCurrentUser() {
    return userService.getCurrentUser();
}
```

This endpoint should expose the currently logged-in user's information. It's delegating the call to the userService.getCurrentUser() method. However, we didn't implement this logic yet, and currently, it will return the user with the lowest ID available in the database. Note that this time, the @Path annotation uses a fixed String literal, so the endpoint should be available at the /api/v1/users/self URL.

We've completely analyzed the UserResource implementation. However, there's still a problem left when deserializing User entities from JSON; let us now see how to fix that.

Deserializing the User entity's password field

When we initially implemented the User entity, we declared all of its fields as public except for the password field. We did this to prevent this field from being exposed through the HTTP API. However, since we are using the same User entity class to deserialize HTTP request bodies containing the User information, there is no way for Quarkus to set the data for this field.

To fix this, we just need to add the following snippet and the required imports to the User class:

```
@JsonProperty("password")
public void setPassword(String password) {
    this.password = password;
}
```

The method will be used by Jackson to deserialize the value of the password property. However, since the password instance variable is still package-private, it won't be serialized.

Let's now repeat the cURL command invocation to create a user:

```
curl -X POST -d"{\"name\":\"packt\",\"password\":\"pass\"}" -H
"Content-Type: application/json" localhost:8080/api/v1/users
```

We should be able to see the following success message:

$ curl -X POST -d'{"name":"packt","password":"pass"}' -H "Content-Type: application/json" localhost:8080/api/v1/users
{"id":2,"name":"packt","created":"2022-04-03T13:42:22.255684853+02:00","version":0,"roles":null}

Figure 3.7 – A screenshot of the result of executing cURL to create a new user

We have now completed the implementation of the HTTP API to deal with the application's users. Let us now continue by implementing the API to manage the user's projects.

ProjectResource

This resource will expose the public operations of the `ProjectService` class, allowing the HTTP API's logged-in users to perform actions concerning their task manager projects. We'll start by creating a new `ProjectResource` class in the `com.example.fullstack.project` package.

The implementation of the `ProjectResource` class is very similar to the one for the `UserResource` class – you can find the complete code in the GitHub repository at `https://github.com/PacktPublishing/Full-Stack-Quarkus-and-React/blob/main/chapter-03/src/main/java/com/example/fullstack/project/ProjectResource.java`. Just as we did with the `UserResource` class, we'll annotate the `ProjectResource` class with the `@Path` annotation – yet, in this case, we'll expose it under the `/api/v1/projects` URL. Again, all of the operations will be delegated to the service class that encapsulates the business logic: `ProjectService`. An instance of this class is injected so that it can be reused in the method implementations. The class contains most of the methods and annotations we implemented for `UserResource`: `get`, `create`, `update`, and `delete`. The same explanations apply here.

If the application is running, once we save the new class, we can start to experiment with the new endpoints.

> **Note**
> We didn't implement the application's security yet, so every project operation will apply to the user with the lowest ID available in the database.

- We can start by creating a new project by executing the following command:

```
curl -X POST -d"{\"name\":\"project\"}" -H "Content-Type:
application/json" localhost:8080/api/v1/projects
```

The execution should complete and print a JSON object with the details of the new project. The JSON should look something similar to the following (note that we've omitted and truncated some fields to make it more legible):

```
{"id":10,"name":"project","user":{"id":0…},"version":0}
```

- We should also be able to list the user's projects by executing the following command:

```
curl localhost:8080/api/v1/projects
```

The command should complete and print a JSON object with the list of projects for the user, including the one we just created:

```
[{"id":10,"name":"project","user":{"id":0…},"version":0}]
```

- We can also update the project and change its name by executing the following cURL command – note that the URL should be adapted to include the id that was returned in the previous create command:

```
curl -X PUT -d"{\"name\":\"new-name\",\"version\":0,
\"user\":{\"id\":0}}" -H "Content-Type: application/json"
localhost:8080/api/v1/projects/10
```

The command should print a JSON object with the updated project information:

```
{"id":10,"name":"new-name","user":
{"id":0…},"version":1}
```

- We should also be allowed to delete the newly created project by executing the following command:

```
curl -X DELETE localhost:8080/api/v1/projects/10
```

Since the endpoint doesn't produce a response body, the command will just complete successfully.

We've covered the application's HTTP API for users and projects; now, let's complete it by implementing the TaskResource class.

TaskResource

This resource will expose the public operations of the TaskService class through an HTTP API. It allows logged-in users to perform CRUD operations on their tasks and mark them as complete. We'll start by creating a new TaskResource class in the com.example.fullstack.task package.

The implementation of the TaskResource class is very similar to those for the UserResource and ProjectResource classes – you can find the complete code in the GitHub repository at https://github.com/PacktPublishing/Full-Stack-Quarkus-and-React/blob/main/chapter-03/src/main/java/com/example/fullstack/task/TaskResource.java. Just as we did for the other resource classes, we'll annotate the class with the @Path annotation and expose it under the /api/v1/tasks URL. We will also inject an instance of the TaskService class. This class contains the get, create, update, and delete methods, which have implementations and annotations that are almost identical to the ones in the UserResource and ProjectResource classes covered previously.

The class contains an additional method: public Uni<Boolean> setComplete(@PathParam("id") long id, boolean complete). This endpoint allows users to mark tasks as complete. It's annotated with @PUT and exposed under the /api/v1/tasks/{id}/complete URL where {id} is a parameter to be replaced with the project ID.

Once we save the new class, if the application is running via the `./mvnw quarkus:dev` command, we should be able to test some of the new endpoints:

- The following `curl` command could be used to create a new task:

```
curl -X POST -d"{\"title\":\"task\"}" -H "Content-Type:
application/json" localhost:8080/api/v1/tasks
```

The execution should complete and print a JSON object with the details of the new task. The JSON should look something similar to the following (note that we've omitted and truncated some fields to make it more legible):

```
{"id":11,"title":"task","description":null,"priority":nu
ll,"user":{"id":0...},"complete":null,"project":null,"vers
ion":0}
```

- You can now mark this task as complete by sending the following HTTP request (note that the URL should be adapted to include the `id` that was returned in the previous `create` command):

```
curl -X PUT -d"\"true\"" -H "Content-Type:
application/json" localhost:8080/api/v1/tasks/11/complete
```

The command should complete successfully and print `true` on the screen.

You should be able to test the rest of the endpoints by yourself by adapting the cURL commands we described for projects.

I would like to highlight how we're propagating the *asynchronous* `Uni` return type across the different classes and taking full advantage of the non-blocking reactive capabilities offered by Quarkus. From the initial query to the database, through the business logic and data processing, down to the JAX-RS endpoint definition: everything is encapsulated within an asynchronous Mutiny pipeline.

We have now completed the implementation of all the HTTP endpoints for our application. However, you might have already noticed that if some exception happens, a nasty error message is printed on the screen. Let us now see how to deal with these exceptions and prepare proper HTTP responses.

Dealing with service exceptions

To be able to handle the application's exceptions and map them to proper HTTP responses, we need to provide an implementation of `ExceptionMapper`. We will start by creating a new `RestExceptionHandler` class in the `com.example.fullstack` package. In the following code snippet, you will find the relevant source code for the class declaration (you can find the complete code in the GitHub repository at `https://github.com/PacktPublishing/Full-Stack-Quarkus-and-React/blob/main/chapter-03/src/main/java/com/example/fullstack/RestExceptionHandler.java`):

```
@Provider
public class RestExceptionHandler implements
```

```
ExceptionMapper<HibernateException> {
  // ...
}
```

The JAX-RS specification defines the `ExceptionMapper` interface to be able to customize the way Java exceptions are converted to HTTP responses. To write a custom `ExceptionMapper`, we need to create a class that implements this interface and annotates it with the `@Provider` annotation so that it is automatically discovered by Quarkus. In this case, we are implementing it with a `HibernateException` type parameter. This means that only exceptions that extend `HibernateException` will be processed by this mapper, which should be fine for our application.

Let us now implement and analyze the methods for this class:

- `hasExceptionInChain(...)`:

```
private static boolean hasExceptionInChain(
  Throwable throwable, Class<? extends Throwable>
    exceptionClass) {
  return getExceptionInChain(throwable,
    exceptionClass).isPresent();
}
```

This method checks whether the current exception, `throwable`, or any of the exceptions in its stack trace are an instance of the provided `exceptionClass`:

- `hasPostgresErrorCode(Throwable throwable, String code)`

```
private static boolean hasPostgresErrorCode(
  Throwable throwable, String code) {
  return getExceptionInChain(throwable,
    PgException.class)
    .filter(ex -> Objects.equals(ex.getCode(), code))
    .isPresent();
}
```

This method tries to retrieve a `PgException` from the provided `throwable` exception's stack trace. If a `PgException` is found, it then checks to see whether it contains the provided error code. We will initially use this method to identify when a database's unique constraint has been violated.

- `toResponse(HibernateException exception)`:

```
public Response toResponse(HibernateException
  exception) {
  if (hasExceptionInChain(exception,
```

```
        ObjectNotFoundException.class)) {
        return Response.status(Response.Status.NOT_FOUND)
          .entity(exception.getMessage()).build();
    }
    if (hasExceptionInChain(exception,
      StaleObjectStateException.class)
      || hasPostgresErrorCode(exception,
        PG_UNIQUE_VIOLATION_ERROR)) {
      return Response.status
        (Response.Status.CONFLICT).build();
    }
    return Response.status(Response.Status.BAD_REQUEST)
      .entity("\"" + exception.getMessage() + "\"").
        build();
}
```

This method contains the logic to effectively map the exceptions into HTTP responses with more suitable HTTP status codes. If the exception corresponds to ObjectNotFoundException, a response with a 404 Not Found status code will be returned. If the exception is of the StaleObjectStateException type, which is thrown when there is an optimistic lock problem, a 409 Conflict status code will be returned instead. The same 409 Conflict status code will be provided if the exception corresponds to a PostgreSQL unique key violation. In any other case, a response with a standard 400 Bad Request status code will be returned.

This method could be improved in the future to cover other kinds of exceptions or to provide more details in the response. In addition, you could also implement an additional ExceptionMapper to deal with other kinds of exceptions too.

We can now execute the application in dev mode using the ./mvnw quarkus:dev command and check whether our new RestExceptionHandler is working.

We can start by performing a request to query a user that doesn't exist:

```
curl -i localhost:8080/api/v1/users/1337
```

Note that we've included the -i command-line flag to print the response headers and body. You should be able to see something similar to the following:

Figure 3.8 – A screenshot of a cURL execution showing the 404 Not Found status

We can also force a conflict error by issuing a request to create a duplicate user:

```
curl -i -X POST -d"{\"name\":\"admin\",\"password\":\"pass\"}"
-H "Content-Type: application/json" localhost:8080/api/v1/users
```

Once executed, the following headers should be visible:

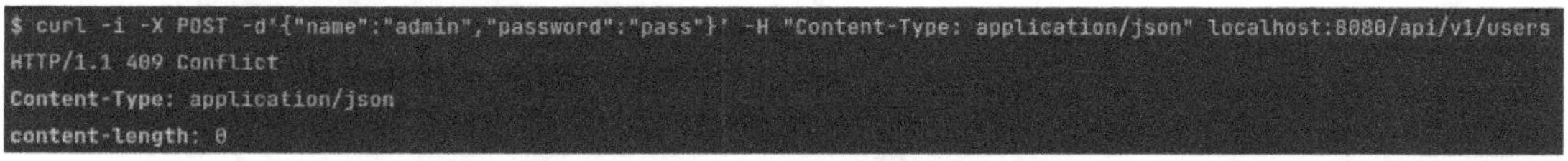

Figure 3.9 – A screenshot of a cURL execution showing the 409 Conflict status

We've now implemented a common exception mapper that will convert Java exceptions to HTTP responses. This is very useful for the frontend side of the application, which can now properly handle any of the exceptions managed by our `RestExceptionHandler`.

Summary

In this chapter, we learned how to implement both blocking and non-blocking endpoints in Quarkus using RESTEasy Reactive. We also implemented the complete business logic and HTTP API for our application. We started by developing the business logic for the task manager in different service classes. Then, we implemented the JAX-RS endpoints in resource controller classes that receive these services via dependency injection. We also learned how to map Java exceptions to HTTP responses to be able to provide more accurate response status codes and how to fully customize the response.

You should now be able to implement HTTP and REST APIs in Quarkus. In the next chapter, we'll see how to secure the application using JWT. We'll implement JWT authentication and authorization and protect the endpoints we just developed.

Questions

1. Can RESTEasy Reactive be used to implement synchronous, blocking endpoints?

2. What are the two types provided by Mutiny to start a pipeline?

3. What is a bean?

4. How can you easily declare a singleton bean in Quarkus?

5. Why is bcrypt preferred for hashing passwords?

6. How can you add a path parameter to a URL in Quarkus?

7. Is it necessary to include the @Produces JAX-RS annotation in Quarkus endpoint definitions?

8. How can you intercept a Java exception and map it to an HTTP response?

4

Securing the Application

In this chapter, we'll implement a security layer based on **JSON Web Token (JWT)**, pronounced "jot," to protect the HTTP API we developed in *Chapter 3, Creating the HTTP API*. We'll start by learning about Quarkus security and its JWT-related extensions. We will also add the required dependencies and learn about what each of them provides. After that, we'll implement the security for our task manager application. We'll generate and configure the required key files, create an authentication and authorization service, and protect the HTTP endpoints.

By the end of this chapter, you should be able to provide a security layer for your Quarkus application based on JWT. You should also have a basic understanding of the Quarkus security module and how to generate your own JWTs.

We will be covering the following topics in this chapter:

- Using JWT security in Quarkus
- Implementing the task manager's HTTP API security

Technical requirements

You will need the latest Java JDK LTS version (at the time of writing, Java 17). In this book, we will be using Fedora Linux, but you can use Windows or macOS as well.

You will need a working Docker environment to take advantage of Quarkus Dev Services. There are Docker packages available for most Linux distributions. If you are on a Windows or macOS machine, you can install Docker Desktop.

If you're not using IntelliJ IDEA Ultimate edition, you'll need a tool such as **cURL** or **Postman** to interact with the implemented HTTP endpoints.

You will need the OpenSSL command-line tool or an alternative to be able to generate the JWT signing keys.

You can download the full source code for this chapter from `https://github.com/PacktPublishing/Full-Stack-Quarkus-and-React/tree/main/chapter-04`.

Using JWT security in Quarkus

Quarkus has an extensive list of modules and extensions to provide a security layer for your application. **Quarkus Security** is the base module upon which the rest of the security extensions are built. You can choose from many of the supported authentication and authorization mechanisms: basic authentication, OpenID Connect, OAuth2, JWT, and so on. On top of that, Quarkus Security provides additional tools to improve the developer experience and testing utilities to enhance the overall application quality and reliability.

In this book, we'll cover how to implement a JWT-based security authentication and authorization mechanism for Quarkus applications. In the *What is Quarkus?* section in *Chapter 1, Bootstrapping the Project*, we learned that Quarkus is based on a set of proven standards and libraries. This is also the case for JWT, which is based on the **MicroProfile JWT RBAC** security specification of the JWT standard and is provided by the **SmallRye JWT** implementation.

> JWT
>
> **JSON Web Token (JWT)** is an **Internet Engineering Task Force (IETF)**-proposed standard that can be used to represent and securely exchange claims between two parties. It is now one of the most widely adopted methods for authorization and information exchange, especially in microservice and distributed architectures, due to its self-contained and compact nature. A token is just a JSON object that contains a set of claims and is usually digitally signed or protected with a **Message Authentication Code (MAC)**.

In general, you'll want an external identity manager service such as **Keycloak** to perform the authentication and generate the JWTs for your application. In this case, you'd only need to configure your application's *authorization* to verify the bearer token upon any received HTTP request to a secured endpoint. However, for the sake of completeness, in our task manager application, we'll also implement an *authentication* mechanism that will generate JWTs too. Let us now see what the required dependencies are to support the generation and validation of JWTs in our application.

Adding the required dependencies

We will start by adding all the required dependencies and then analyze what each of the dependencies provides. Just like we did in the previous chapters, we'll use the `quarkus:add-extension` Maven goal, which will automatically insert the required entries in the project's `pom.xml`. You can now execute the following command in the project's root to add the dependencies:

```
./mvnw quarkus:add-extension -Dextensions=smallrye-
jwt,smallrye-jwt-build
```

Once executed, you should see the following message:

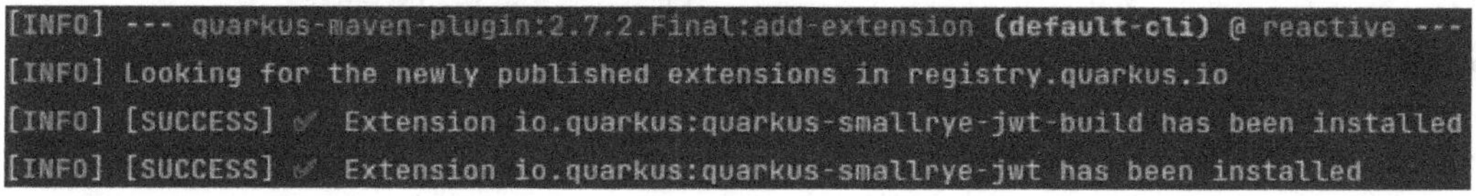

```
[INFO] --- quarkus-maven-plugin:2.7.2.Final:add-extension (default-cli) @ reactive ---
[INFO] Looking for the newly published extensions in registry.quarkus.io
[INFO] [SUCCESS] ✔ Extension io.quarkus:quarkus-smallrye-jwt-build has been installed
[INFO] [SUCCESS] ✔ Extension io.quarkus:quarkus-smallrye-jwt has been installed
```

Figure 4.1 – A screenshot of the execution result of the quarkus:add-extension command

Let us see what each of these dependencies provides.

SmallRye JWT

The execution of the `quarkus:add-extension` Maven goal should have added the following entry into your `pom.xml`:

```
<dependency>
  <groupId>io.quarkus</groupId>
  <artifactId>quarkus-smallrye-jwt</artifactId>
</dependency>
```

SmallRye JWT is an implementation of the Eclipse MicroProfile JWT RBAC security specification. Quarkus provides its own packaged extension of the SmallRye JWT dependency with build-time optimizations and improvements for native compilation. This dependency is primarily used for the parsing and optional decryption of the JWT and its signature verification.

Internally, Quarkus will automatically and transparently handle the bearer token authorization headers provided in the HTTP requests. It will verify the token's signature, and deserialize it into an `org.eclipse.microprofile.jwt.JsonWebToken` instance. This instance will be used by Quarkus to determine the user's authorized roles and will also be made available for injection.

SmallRye JWT build

After executing the `./mvnw quarkus:add-extension` command, we should also be able to find the following dependency in our `pom.xml`:

```
<dependency>
  <groupId>io.quarkus</groupId>
  <artifactId>quarkus-smallrye-jwt-build</artifactId>
</dependency>
```

The SmallRye JWT build dependency provides us with an straightforward interface to generate a JWT, add some claims to it, and digitally sign it. We can use this to issue trusted JWTs for the users of our application once they've completed an authentication process. The users can, later on, use this token to be authorized when they interact with our application's HTTP API using it as a bearer token in an authorization HTTP request header.

Now that we've learned about JWT and Quarkus security and added the required dependencies to our application, let us see how to implement the security layer.

Implementing the task manager's HTTP API security

You should now be familiar with Quarkus security and JWT, and the project should contain the required dependencies. We can now start to implement and configure the task manager application's security. We'll start by generating the required key files to sign and verify the tokens.

Generating the key files

The JWT standard provides different methods to verify and trust the authenticity of the tokens and the integrity of the claims it contains. One of the most common approaches, and the one that we'll be using in our application, is the usage of signed tokens. In our case, we'll be using a private and public key pair to sign and verify the tokens.

In a distributed application, the authorization service holds the *private* key and uses it to issue the signed JWTs. The rest of the services have access to the *public* key and use it to verify the authenticity of these tokens. In our application, we're going to perform both operations, the signing and issuing of the tokens when we receive authentication requests and their verification when we receive them in an HTTP authorization header.

Before generating the key files, we need to create a directory where we'll store them. To create the directory, we right-click on the `src/main/resources` directory and click on the **New** submenu and the **Directory** menu entry.

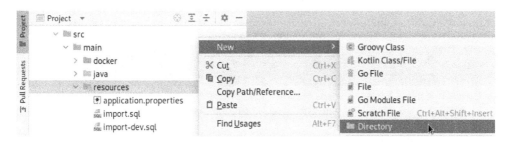

Figure 4.2 – A screenshot of the IntelliJ File | New | Directory menu entry

In the modal dialog form, we will type the name of the new directory: jwt.

New Directory

jwt

Figure 4.3 – A screenshot of the IntelliJ New Directory dialog

We can now navigate to the new `src/main/resources/jwt` directory in a terminal and issue the following command to generate the private key:

```
openssl genrsa -out rsa-private-key.pem 2048
```

The command should complete successfully and a message like the following should be printed:

```
$ openssl genrsa -out rsa-private-key.pem 2048
Generating RSA private key, 2048 bit long modulus (2 primes)
.............................+++++
..+++++
e is 65537 (0x010001)
```

Figure 4.4 – A screenshot of the result of the command to generate the private key

Next, we need to convert the private key into `PKCS#8` format to make it compatible with the SmallRye JWT build. We can achieve this by executing the following command:

```
openssl pkcs8 -topk8 -nocrypt -inform pem -in rsa-private-key.
pem -outform pem -out private-key.pem
```

The command should complete successfully without printing any additional message.

The final step in the key generation phase would be to create the public key to verify the signed tokens. We need to execute the following command:

```
openssl rsa -pubout -in rsa-private-key.pem -out public-key.pem
```

The command should complete successfully and the following message should be printed:

```
$ openssl rsa -pubout -in rsa-private-key.pem -out public-key.pem
writing RSA key
```

Figure 4.5 – A screenshot of the result of the command to generate the public key

We have created the required key files to sign and verify the tokens. Let us now configure the application to use these files.

Configuring the application

SmallRye JWT needs some configurations defined to be able to load the newly generated key files. We need to add the following entries to the `application.properties` file:

```
smallrye.jwt.sign.key.location=jwt/private-key.pem
mp.jwt.verify.publickey.location=jwt/public-key.pem
mp.jwt.verify.issuer=https://example.com/issuer
```

Let us see what each of these configuration properties does:

- `smallrye.jwt.sign.key.location`

  This is the property to define the location of the *private* key that will be used by SmallRye to sign the tokens issued from our authorization service. We'll set the path of the PKCS #8-formatted private key we just generated.

- `mp.jwt.verify.publickey.location`

  This is the property to define the location of the *public* key that will be used to verify the signatures of JWTs. For our application, we are going to set the local path of the public key we generated. However, this property can also point to a remote URL. This would be the case if you were using an external authorization service such as Keycloak.

- `mp.jwt.verify.issuer`

  This is the configuration property to define the value to verify the `iss` (issuer) claim that should be included within the JWT. Like the public key location, this value can point to a remote URL in case our application uses an external authorization service.

The resulting `application.properties` file should now look like this:

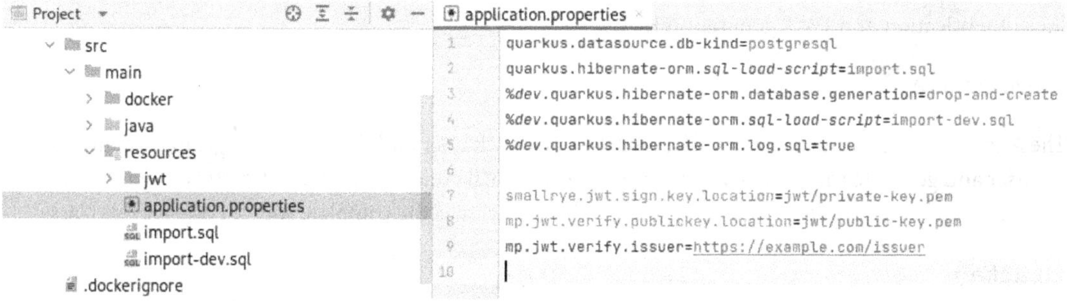

```
application.properties
1   quarkus.datasource.db-kind=postgresql
2   quarkus.hibernate-orm.sql-load-script=import.sql
3   %dev.quarkus.hibernate-orm.database.generation=drop-and-create
4   %dev.quarkus.hibernate-orm.sql-load-script=import-dev.sql
5   %dev.quarkus.hibernate-orm.log.sql=true
6
7   smallrye.jwt.sign.key.location=jwt/private-key.pem
8   mp.jwt.verify.publickey.location=jwt/public-key.pem
9   mp.jwt.verify.issuer=https://example.com/issuer
10
```

Figure 4.6 – A screenshot of the configured properties in the application.properties file

We have configured the application to make use of the generated keys. Let us now continue with the implementation of the authentication service.

Implementing the authentication service and login interface

Users will need to authenticate and obtain a valid JWT with the required claims to be able to consume our application's HTTP API once it's protected. We are now going to implement an authentication service to provide this functionality.

However, we'll first add a utility method to the `UserService` class to verify that an authentication request password matches the one in the database. The next snippet contains the code we need to add to the `UserService` class:

```
public static boolean matches(User user, String password) {
    return BcryptUtil.matches(password, user.password);
}
```

This method checks that the provided *plain text* password matches the Modular Crypt-formatted *bcrypt* hashed password stored in the also provided `User` entity. The method returns `true` if the passwords match; else, it returns `false`.

We can now create the authorization service. First, we'll create a new `com.example.fullstack.auth` package where we'll host the related classes. In this package, we'll add a new `AuthRequest` record that we'll use to hold the login credentials provided by users logging into the application. In the following snippet, you will find the complete source code for this record:

```
package com.example.fullstack.auth;
public record AuthRequest(String name, String password) {
}
```

It's a very simple record that contains two fields: name and password. The record will be used to deserialize authentication requests. Let us now implement the AuthService class, which will hold the authentication and JWT generation logic.

AuthService

The AuthService class is the core class for the current chapter and contains the logic to authenticate the user and generate the JWTs with the user's claims. We'll create a new AuthService class in the com.example.fullstack.auth package; the following is the complete source code for this class:

```java
package com.example.fullstack.auth;
import com.example.fullstack.user.UserService;
import io.quarkus.security.AuthenticationFailedException;
import io.smallrye.jwt.build.Jwt;
import io.smallrye.mutiny.Uni;
import org.eclipse.microprofile.config.inject.
    ConfigProperty;
import javax.enterprise.context.ApplicationScoped;
import javax.inject.Inject;
import java.time.Duration;
import java.util.HashSet;


@ApplicationScoped
public class AuthService {

  private final String issuer;
  private final UserService userService;
  @Inject
  public AuthService(
    @ConfigProperty(name = "mp.jwt.verify.issuer") String
      issuer, UserService userService) {
    this.issuer = issuer;
    this.userService = userService;
  }
  public Uni<String> authenticate(AuthRequest authRequest)
    {
    return userService.findByName(authRequest.name())
      .onItem()
```

```
        .transform(user -> {
          if (user == null || !UserService.matches(user,
             authRequest.password())) {
            throw new AuthenticationFailedException
             ("Invalid credentials");
          }
          return Jwt.issuer(issuer)
             .upn(user.name)
             .groups(new HashSet<>(user.roles))
             .expiresIn(Duration.ofHours(1L))
             .sign();
        });
    }
  }
```

Just like we did for the rest of the task manager application services, we'll annotate this class with the @ApplicationScoped annotation to define a *singleton* bean. The class constructor is annotated with the @Inject annotation to take advantage of constructor-based injection. In this case, the class has two dependencies: userService and issuer. Now, UserService will be used to retrieve the user information to complete the JWT. The issuer string parameter is annotated with the @ConfigProperty annotation. This annotation instructs Quarkus to inject a configuration value (in this case, mp.jwt.verify.issuer), which we provided in the application. properties file. The mp.jwt.verify.issuer property is used by SmallRye to verify the token issuer. However, since the value is available from the configuration, we'll reuse it here for the token generation too.

The class contains a single method, authenticate, with a single parameter, AuthRequest. The method returns a Uni that emits an item containing the signed JWT or a failure if the user is not found or the password doesn't match. The implementation starts by trying to find the requested user by name. The user is then checked asynchronously; if the user doesn't exist (is null) or the passwords don't match, an AuthenticationFailedException instance is thrown and will be emitted as a failure.

If the user exists and the passwords match, a JWT is created and signed using the SmallRye Jwt builder class. The JWT is built with the issuer (iss), the user principal name (upn), and the groups claim. The groups claim contains the user roles and will be used later on to check whether the user is authorized to perform any given operation.

We are also setting the expiration time for the token. In this case, we're setting the token to expire after 1 hour of its creation. This means that the token will no longer be valid after this period and users will need to reauthenticate. In general, we would want to set a shorter lifetime and provide an additional *refresh token*. This token could be used by the users, or more specifically the frontend application, to transparently generate a new JWT without having to perform the login operation again. For this book, we won't be covering this part, since from the backend perspective, it would be a similar operation to the one we just performed. You also have all of this functionality out of the box if you delegate the authentication services to a production-grade system such as Keycloak.

We have now covered the logic to authenticate the user and issue a JWT. Next, we will see how to consume the exposed service through the HTTP API.

AuthResource

The AuthResource class exposes an HTTP endpoint to enable users to authenticate and consume the AuthService authenticate method. Just like we did for the AuthService class, we'll create a new AuthResource class in the com.example.fullstack.auth package. The following code snippet contains the source code for this class:

```
package com.example.fullstack.auth;
import io.smallrye.mutiny.Uni;
import javax.annotation.security.PermitAll;
import javax.inject.Inject;
import javax.ws.rs.POST;
import javax.ws.rs.Path;
@Path("/api/v1/auth")
public class AuthResource {
  private final AuthService authService;
  @Inject
  public AuthResource(AuthService authService) {
    this.authService = authService;
  }
  @PermitAll
  @POST
  @Path("/login")
  public Uni<String> login(AuthRequest request) {
    return authService.authenticate(request);
  }
}
```

The class is annotated with the @Path annotation to expose all of the authentication-related endpoints with the /api/v1/auth prefix. The AuthService bean instance is injected into the class via constructor-based injection so that it can be reused in the latter method.

The login(AuthRequest request) method exposes an endpoint that allows users to perform login HTTP POST requests at the /api/v1/auth/login path. This method delegates the call to the authenticate method in AuthService, which we already covered.

Notice that the method is annotated with an @PermitAll annotation, which we haven't seen before. This annotation specifies that the endpoint can be consumed by *any* user regardless of its assigned roles. Since this is the authentication entry point for all users, it makes sense to keep it public and unchecked.

If we start the application via ./mvnw quarkus:dev, we should be able to test whether the endpoint works and generates JWTs. We can execute the following cURL command to retrieve a valid token:

```
curl -X POST -d'{"name":"user","password":"quarkus"}' -H
"Content-Type: application/json" localhost:8080/api/v1/auth/
login
```

If everything goes well, a new token should be printed:

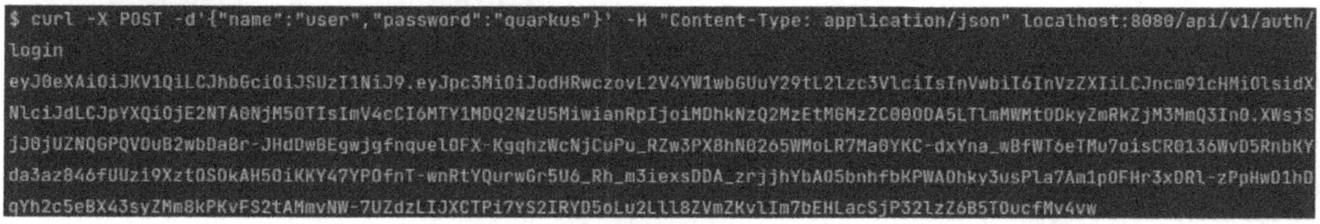

Figure 4.7 – A screenshot of the result of executing cURL to perform a valid login

---

cURL on Windows cmd.exe

If you are using cURL on Windows within the cmd.exe terminal, you might need to use double quotes instead of single quotes, and escape the ones within the data field. For example, the previous command should be invoked as curl -X POST -d"{""name"":""user"",""password"":""quarkus""}" -H "Content-Type: application/json" localhost:8080/api/v1/auth/login.

We can also check what happens if we provide an invalid password by executing the following command (note the -i flag to print the response headers):

```
curl -X POST -d'{"name":"user","password":"invalid"}' -H
"Content-Type: application/json" -i localhost:8080/api/v1/auth/
login
```

We should now be able to see the unauthorized response code:

```
$ curl -X POST -d'{"name":"user","password":"invalid"}' -H "Content-Type: application/json" -i localhost:8080/api/v1/au
th/login
HTTP/1.1 401 Unauthorized
Content-Type: text/plain;charset=UTF-8
www-authenticate: Bearer
content-length: 0
```

Figure 4.8 – A screenshot of the result of executing cURL to perform an invalid login

We have now implemented the authorization service and enabled users to log into the application. Let us now see how to get the currently logged-in user to be able to perform the user-specific operations.

Getting the logged-in user

In the *UserService* section in *Chapter 3*, *Creating the HTTP API*, we created a dummy implementation of the getCurrentUser method. Now that users can perform login operations and use the resulting tokens to provide a bearer token authorization header, we can provide the actual implementation.

We'll start by adding an instance variable and constructor to the UserService class:

```
private final JsonWebToken jwt;
@Inject
public UserService(JsonWebToken jwt) {
   this.jwt = jwt;
}
```

This code instructs Quarkus to inject the JWT as a JsonWebToken instance variable to the UserService class bean. If the HTTP request contained a bearer token authorization header, Quarkus security will parse and verify the token.

> **Note**
> Despite injecting the JsonWebToken instance to an @ApplicationScoped annotated *singleton* bean, Quarkus will inject the actual JWT used in the *current request*.

We can now replace the getCurrentUser method with the following implementation:

```
public Uni<User> getCurrentUser() {
   return findByName(jwt.getName());
}
```

The method just retrieves the name from the Principal, which in this case represents a user, and tries to find a matching user in the database. With this new implementation, all of the user-scoped operations we implemented in the project and task services will now work as expected.

Now that we can retrieve the currently logged-in user, let us further improve the user service to allow for a specific password-change operation.

Letting users change their password

In the *UserService* section in *Chapter 3, Creating the HTTP API*, we provided an initial implementation of the update method. This implementation allows *any* authorized user to perform a complete modification of any of the user's entity fields. Since we can now retrieve the current user, it makes sense to provide a password-change operation and restrict the update method to only allow modifications of some of the entity fields.

We'll start by replacing the update method implementation with the following:

```
@ReactiveTransactional
public Uni<User> update(User user) {
   return findById(user.id).chain(u -> {
      user.setPassword(u.password);
      return User.getSession();
   })
   .chain(s -> s.merge(user));
}
```

We've added the user.setPassword(u.password); instruction to the chained lambda expression. This will overwrite any password value provided by the user's request with the one in the database, preventing the current password from being modified.

We'll now implement a specific changePassword method in the UserService class; the following snippet contains the code for the implementation:

```
public Uni<User> changePassword(String currentPassword,
   String newPassword) {
   return getCurrentUser()
      .chain(u -> {
         if (!matches(u, currentPassword)) {
            throw new ClientErrorException("Current password
               does not match", Response.Status.CONFLICT);
         }
         u.setPassword(BcryptUtil.bcryptHash(newPassword));
         return u.persistAndFlush();
      });
}
```

The method performs a write operation in the database. On this basis, we annotate it with the `@ReactiveTransactional` annotation to configure Quarkus to run the executed method in a reactive Mutiny transaction for its persistence.

The method's signature contains two parameters: `currentPassword` and `newPassword`. The method returns a `Uni` that will emit an item with the modified `User` entity or a failure if the provided `currentPassword` doesn't match the user's password in the database.

The implementation starts by retrieving the currently logged-in user's entity. It then checks whether the `currentPassword` argument matches the one in the database. If it doesn't, `ClientErrorException` is thrown. When this exception is processed by Quarkus, a `409 Conflict` status code will be returned in the HTTP response. If the password matches, the `newPassword` argument is hashed and set in the user entity, which is finally persisted in the database.

We can now expose the password-change operation in the HTTP API. However, we'll first create a new `PasswordChange` record in the `com.example.fullstack.user` package with the following content:

```
package com.example.fullstack.user;


public record PasswordChange(String currentPassword, String
    newPassword) {

}
```

The record contains two fields: `currentPassword` and `newPassword`. The record will be used to deserialize the HTTP body of password-change requests.

We can now add a new `changePassword` method to the `UserResource` class to expose the password-change operation. The following code snippet contains the required code:

```
@PUT
@Path("self/password")
@RolesAllowed("user")
public Uni<User> changePassword(PasswordChange
    passwordChange) {
    return userService
        .changePassword(passwordChange.currentPassword(),
            passwordChange.newPassword());
}
```

The method is annotated with an `@PUT` operation, which indicates that this method should be used to respond to HTTP PUT requests. The method is also annotated with an `@Path` annotation with a `self/password` value; the resulting endpoint will be available at the `/api/v1/users/self/password` URL.

We covered most of these annotations before; however, note the new `@RolesAllowed` annotation. This annotation instructs Quarkus to restrict this endpoint to users who have the `user` role. If you recall, in the *Loading the application's initial data* section in *Chapter 2*, *Adding Persistence*, we created two users: `admin` and `user`, both with the `user` role. In this case, both users would be able to invoke this method.

The method implementation delegates the call to the `UserService changePassword` method we just added.

If we start the application via `./mvnw quarkus:dev`, we should be able to test whether the endpoint works and be able to change the user's password. However, we first need a valid JWT; for this, we can use the cURL command we executed in the *AuthResource* section. Next, we can execute the following cURL command to change the password (you need to replace the `$jwtToken` with the one you retrieved):

```
curl -X PUT
-d'{"currentPassword":"quarkus","newPassword":"changed"}' -H
"Content-Type: application/json" -H "Authorization: Bearer
$jwtToken" localhost:8080/api/v1/users/self/password
```

If everything goes well, you should be able to see a message similar to this one:

```
$ curl -X PUT -d'{"currentPassword":"quarkus","newPassword":"changed"}' -H "Content-Type: application/json"
-H "Authori
zation: Bearer eyJ0eXAiOiJKV1QiLCJhbGciOiJSUzI1NiJ9.eyJpc3MiOiJodHRwczovL2V4YW1wbGUuY29tL2lzc3VlciIsInVwbiI6
InVzZXIiLCJncm91cHMiOlsidXNlciJdLCJpYXQiOjE2NTA0NzIyNTIsImV4cCI6MTY1MDQ3NTg1MiwianRpIjoiZDI1NDUxOTktZTE3NC00
NTk2LWJiMzQtOWE5MDllMmZiNTg2In0.VSfNoDx5NQg4rY1fn1MT7AnwtHRZKD7VAKZSehn 0BMVUurt8aRLQBLgHtUU8YPffZhlk6CArapVu
yKGbeVRiPLW_6lKgfuacdDetjJqDWv580Q5FPh4QF7esnwepT07TCZDorMPfPtRWzeRefeIZH2Af7DRgyDyrFsKeua5Hu6te5FnOB4LKvYuS
QaNiNZC6Ek_t8r7lqCReUVQRIiRlME_lsdLYh6wKnCHvSBVcc-YI2YixbRqXT8NlFnprSrJAP6St4Kzqgv0RO7Zdb2x0vLT6FCm2R-8qz7uB
KI60yYCaxcmN4B3-FIzWVvFjCG4HAG42G3DO4cHhtvM0Kg6_BA" localhost:8080/api/v1/users/self/password
{"id":1,"name":"user","created":"2022-04-20T16:29:37.133359+02:00","version":1,"roles":["user"]}
```

Figure 4.9 – A screenshot of the result of executing cURL to change the user password

If you repeat the command execution without any changes, it should fail since the password was modified and `currentPassword` won't match the one in the database. In *Chapter 5*, *Testing Your Backend*, we'll implement integration tests to verify all of these scenarios.

We've refactored the user entity update mechanism and provided and exposed a specific operation for users to change their passwords. Let us now see how to secure the rest of the application's endpoints.

Securing the HTTP resources

Now that we have an authentication service in place and we've configured the application to use JWT authorization, we can secure the HTTP exposed resources. In the *Letting users change their password* section, we learned about the @RolesAllowed annotation and used it to annotate a method and restrict its invocation to users with the user role. This annotation can be specified at both a class level and a method level. When specified at a class level, it applies to all of the methods in that class. If the method is applied both at a class level and then in a specific method, the method annotation takes precedence over the other.

Let us annotate each of the resource classes and specific methods:

- ProjectResource

  This class exposes endpoints to manage the user's projects. We'll annotate the class with @RolesAllowed("user") to only allow logged-in users with the user role to operate.

- TaskResource

  This class exposes the functionality to manage the user's tasks. Just like we did for the ProjectResource class, we'll annotate it with @RolesAllowed("user").

- UserResource

  This class exposes endpoints to deal mostly with the application's user administrative tasks. In this case, we'll also annotate the class with @RolesAllowed, but we'll configure the admin role instead. The resulting class-level annotation should be @RolesAllowed("admin").

  The getCurrentUser and changePassword methods are meant to be used by regular users. When we implemented the changePassword method, we already annotated the method with a specific @RolesAllowed("user") annotation. We'll need to annotate the getCurrentUser method with this annotation too.

If we start the application via ./mvnw quarkus:dev, we should be able to test whether the annotations work and the endpoints are protected. Let's try to retrieve the list of users by executing the following command:

```
curl -i localhost:8080/api/v1/users
```

We should see an unauthorized response code:

```
$ curl -i localhost:8080/api/v1/users
HTTP/1.1 401 Unauthorized
www-authenticate: Bearer
content-length: 0
```

Figure 4.10 – A screenshot of the unauthorized result of executing cURL to retrieve the list of users

We have now protected the endpoints for the application and implemented a very basic role-based access control layer to protect our application by leveraging our own authorization and authentication service.

Summary

In this chapter, we learned how to implement a security layer in Quarkus using JWT. We learned how to generate and configure private and public key files to sign and verify JWTs. Then, we implemented an authentication service to generate and sign JWTs. We also refactored some of the user-related services and provided functionality for users to change their passwords. We also learned how to use annotations to restrict access to the HTTP API.

You should now be able to secure your Quarkus application by using its JWT extensions and provide role-based access control to specific areas of your application. In the next chapter, we'll provide an introduction and overview of how to test Quarkus applications. We'll also implement integration tests to verify the exposed application functionality.

Questions

1. What is JWT?
2. What key do we need to verify a JWT signature?
3. How can you generate a JWT?
4. Do we need to store a local copy of the configured keys?
5. What annotation can we use to retrieve a configuration value in Quarkus?
6. What takes precedence: an @RolesAllowed class or a method-level annotation?

5

Testing Your Backend

In this chapter, we'll implement tests to verify the task manager's backend features we implemented in the previous chapters. We'll start by learning about testing in Quarkus and its specific dependency to test JWT-secured endpoints. Then, we'll implement and analyze the tests to cover each of the functionalities that we developed for the task manager and learn how to execute them.

By the end of this chapter, you should be able to implement unit and integration tests for your Quarkus applications. You should also have a basic understanding of the Quarkus testing framework. Providing tests for your applications will help you build solid and reliable software and minimize its maintenance effort.

We will be covering the following topics in this chapter:

- Testing in Quarkus
- Testing the task manager

Technical requirements

You will need the latest Java JDK LTS version (at the time of writing, Java 17). In this book, we will be using Fedora Linux, but you can use Windows or macOS as well.

You will need a working Docker environment to take advantage of *Quarkus Dev Services*. There are Docker packages available for most Linux distributions. If you are on a Windows or macOS machine, you can install Docker Desktop.

You can download the full source code for this chapter from `https://github.com/PacktPublishing/Full-Stack-Quarkus-and-React/tree/main/chapter-05`.

Testing in Quarkus

In the *Continuous testing* section in *Chapter 1, Bootstrapping the Project*, we introduced some of Quarkus' testing capabilities. We specifically learned about continuous testing, one of the core testing features that was introduced in Quarkus 2.X. We also examined some of the test code that was bootstrapped from the `code.quarkus.io` wizard and even implemented a new endpoint and its test using **test-driven development (TDD)**.

It's clear that providing a good test framework is one of Quarkus' priorities, including providing the best possible developer experience. In the *Quarkus Dev Services* section in *Chapter 2, Adding Persistence*, we learned about Dev Services and how we don't need to deploy a database when running the application in development mode. Aligned with the goal of providing a great developer experience, Dev Services also works for tests. This means that when Quarkus executes your tests, it will automatically deploy a database for you.

In general, writing a Quarkus unit test is much like writing a regular JUnit 5 unit test. However, there are some nuances that we must be aware of. When you run a Quarkus test, Quarkus starts your application in the background. This makes it ideal to implement tests that integrate several layers of your application. Mocking is still possible via dependency injection, but you need to declare these mocked beans using Quarkus-specific instructions.

> **JUnit**
>
> JUnit is one of the most popular and widespread unit testing frameworks for Java. Developers use this framework to implement and execute automated tests for their Java applications. Its tight integration with almost all IDEs and build tools makes it a perfect candidate for this purpose, and most organizations have adopted it as their main testing framework.

Though most of the required test dependencies were added when we bootstrapped the project, we'll need an additional dependency to be able to test our *secured* application.

Adding the missing test dependency to our project

We'll start by adding the missing dependency for this chapter and then we'll evaluate what this dependency provides. You can execute the following command in the project root and the *Quarkus Maven Plugin* will add the dependency for you:

```
./mvnw quarkus:add-extension -Dextensions=io.quarkus:quarkus-
test-security-jwt
```

Once executed, you should see the following message:

```
[INFO] --- quarkus-maven-plugin:2.7.2.Final:add-extension (default-cli) @ reactive ---
[INFO] [SUCCESS] ✓ Extension io.quarkus:quarkus-test-security-jwt has been installed
```

Figure 5.1 – A screenshot of the execution result of the quarkus:add-extension command

Let us now see what this dependency provides.

Test security JWT

After executing the ./mvnw quarkus:add-extension command, the following dependency should be visible in our pom.xml file. However, we should fine-tune this dependency and set its scope to test:

```
<dependency>
    <groupId>io.quarkus</groupId>
    <artifactId>quarkus-test-security-jwt</artifactId>
    <scope>test</scope>
</dependency>
```

Test security JWT is the dependency provided by Quarkus to test applications that are implemented with a JWT security layer. It has a transitive dependency on the quarkus-test-security artifact, which provides the common test framework used for all of the Quarkus security providers.

Now that we've gone over the Quarkus testing framework and added the required dependencies, let us implement the tests for our task manager application.

Testing the task manager

We've already covered the complete implementation of the task manager backend. In *Chapter 2, Adding Persistence*, we created the *persistence* layer for the application. Then, in *Chapter 3, Creating the HTTP API*, we implemented the *business* and *presentation* layers. Finally, in *Chapter 4, Securing the Application*, we secured the HTTP endpoints of the task manager. We could implement *unit tests* to verify the application functionality for each layer individually. However, this approach would require mocking each layer's dependencies and would require a complex test implementation, which would also mean that its maintenance would be expensive. Since our business logic is not too complex, we'll treat each of the individual application features as a test unit and implement an *integration test* for each functionality covering the three layers. Let us start by configuring our application.

Configuring the application

Whenever we execute a Quarkus test, Quarkus will spin up a real application instance to perform the verifications. Considering this, it's interesting to prepare the test environment and add some initial data to the database. For this purpose, we need to edit the `application.properties` file and add the following entry:

```
%test.quarkus.hibernate-orm.sql-load-script=import-dev.sql
```

In the *Configuring Quarkus* section in *Chapter 2, Adding Persistence*, we configured the same property but for the `dev` profile instead. As we learned, this property defines a path to a file containing SQL statements that will be executed when the application starts. For the testing environment, these statements will be executed when the application starts and before the tests are run. Since our use case and features are quite simple we will reuse the same script that we used for the development environment. This script adds two users, assigns them some roles, and creates a project. If your application grows in complexity or you need to add specific test data, you can create a new file, `import-test.sql`, with your SQL statements and configure this property to point to the new file.

We have now configured the application and it's ready to run in test mode. Let us now implement the tests for the authentication feature.

Testing authentication

In the *Implementing the authentication service and login interface* section in *Chapter 4, Securing the Application*, we developed an authentication service to allow users to log into the application and retrieve a valid JWT. We will now implement a test suite to verify that this feature works as expected.

We'll start by creating a new test suite, `AuthResourceTest`, by opening the `AuthResource` class, right-clicking on its declaration, and clicking on the **Go To** submenu, followed by clicking on the **Test** menu entry:

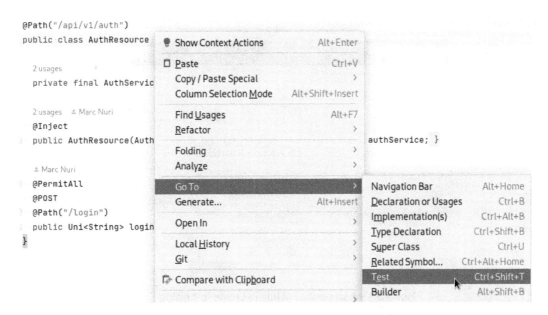

Figure 5.2 – A screenshot of IntelliJ Go To, Test menu entry

We should now see a new pop-up menu where we'll click on the **Create New Test…** entry:

Figure 5.3 – A screenshot of IntelliJ Choose Test pop-up menu

We should now see the following dialog:

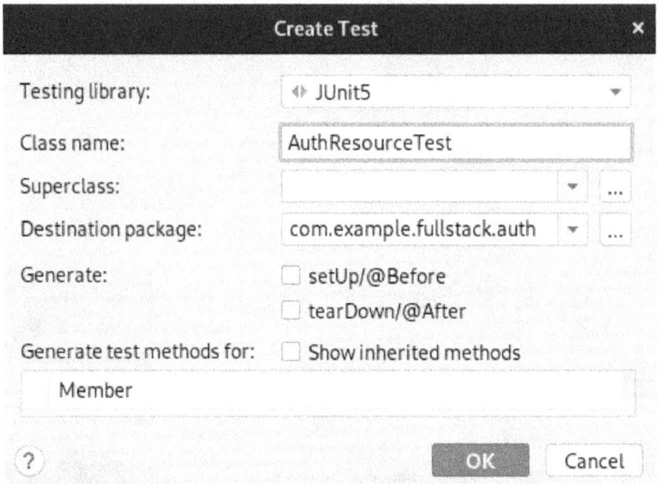

Figure 5.4 – A screenshot of IntelliJ Create Test dialog

The provided values should be fine; we just need to press **OK** and IntelliJ will create the test suite class for us.

In the following code snippet, you will find the relevant source code for the test:

```
@QuarkusTest
class AuthResourceTest {

  @Test
  void loginValidCredentials() {
    given()
       .body("{\"name\":\"admin\",\"password\":\"quarkus\"}")
       .contentType(ContentType.JSON)
       .when().post("/api/v1/auth/login")
       .then()
       .statusCode(200)
       .body(not(emptyString())));
  }

  @Test
  void loginInvalidCredentials() {
    given()
```

```
        .body("{\"name\":\"admin\",\"password\":\
          "not-quarkus\"}")
        .contentType(ContentType.JSON)
        .when().post("/api/v1/auth/login")
        .then()
        .statusCode(401);
   }
}
```

The first thing to notice is that the class is annotated with the `@QuarkusTest` annotation. This configures Quarkus to run the application in test mode so that we can then perform real HTTP requests to verify the application's behavior. We will configure the rest of the test classes with this annotation too.

Let us now examine each of the two test cases defined in the suite:

- `void loginValidCredentials()`

  This test verifies that a user providing *valid* credentials can authenticate and retrieve a valid JWT. The method is annotated with the standard JUnit 5 `@Test` annotation that instructs JUnit to execute this method as a test case. The implementation starts with the `given()` REST Assured **domain-specific language (DSL)** method to build the HTTP request. We'll provide a body for the request with a JSON string, containing valid credentials using the REST Assured `body()` method. These credentials should match those that we added in the `import-dev.sql` script. We'll also provide the `Content-Type` HTTP header for the request, in this case, `application/json`. These invocations have prepared the HTTP request. We can now perform the POST request using the `when()` and `post()` REST Assured DSL methods pointing to the login endpoint.

  REST Assured provides DSL methods to perform assertions for the response of the executed HTTP request too. We start by calling the `then()` method and then verifying that the response has a `200 OK` successful status code and contains a non-empty body.

  Notice how the REST Assured DSL provides methods that help you organize your tests following the common *Given-When-Then* or *Arrange-Act-Assert* testing structure.

- `void loginInvalidCredentials()`

  This test verifies that a user providing *invalid* credentials won't be able to authenticate and retrieve a JWT. The test preparation is very similar to the valid credentials test suite. However, in this case, we'll provide an invalid password for the user. The test assertion now verifies that a `401 Unauthorized` client error status code is returned in the HTTP response.

REST Assured

REST Assured is a library to test and validate REST services in Java. Its support for easily building HTTP requests and its tight integration with Quarkus makes it a great tool to implement integration tests.

Now that we've implemented the test, we can execute it using IntelliJ or through the command line with Maven. To run the test suite in IntelliJ, we need to click on the play button near the class definition and click on the **Run 'AuthResourceTest'** menu entry:

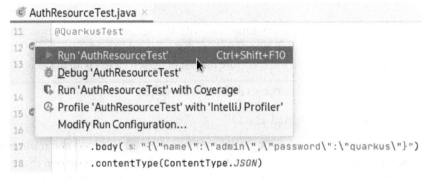

Figure 5.5 – A screenshot of the IntelliJ Run 'AuthResourceTest' menu entry

The tests should execute and pass, and we should be able to see the results in the **Run** tool window:

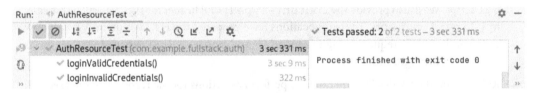

Figure 5.6 – A screenshot of IntelliJ 'AuthResourceTest' test execution results

You can also run the tests with Maven using the standard Maven testing goals:

```
./mvnw clean test
```

Or if you just want to test the current test suite, run the following:

```
./mvnw clean test -Dtest=AuthResourceTest
```

The test should now execute and you should see the Maven build success message:

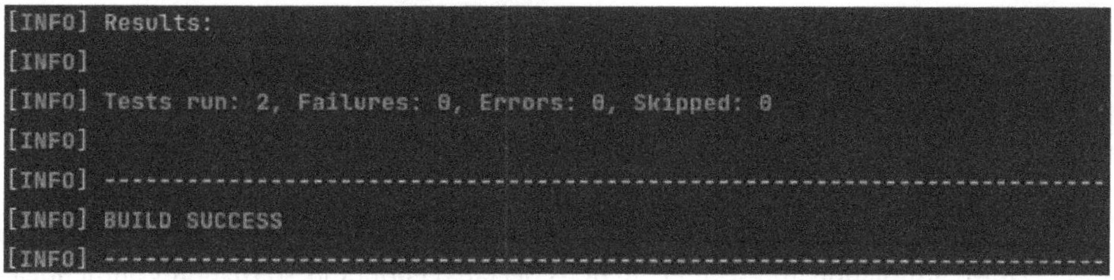

```
[INFO] Results:
[INFO]
[INFO] Tests run: 2, Failures: 0, Errors: 0, Skipped: 0
[INFO]
[INFO] -------------------------------------------------------------
[INFO] BUILD SUCCESS
[INFO] -------------------------------------------------------------
```

Figure 5.7 – A screenshot of the execution result of the mvnw clean
test command for the AuthResourceTest class

Now that we've implemented tests to validate the user authentication features, let us continue by implementing tests for the user-related operations.

Testing the user-related features

In the *UserService* and *UserResource* sections in *Chapter 3, Creating the HTTP API*, we developed and exposed the user-related features. We added endpoints to list, create, update, and delete users, and also two specific endpoints to get the currently logged-in user information and to change the password. We will now implement tests that will verify that all of these features work.

Just like we did for `AuthResource`, we'll open the `UserResource` class, right-click on its declaration, and click on the **Go To** submenu, and the **Test** menu entry. We'll click on the **Create New Test** popup, and accept the default dialog settings. IntelliJ should now create and open a new `UserResourceTest` class.

You can find the full source code for the test suite at `https://github.com/PacktPublishing/Full-Stack-Development-with-Quarkus-and-React/tree/main/chapter-05/src/test/java/com/example/fullstack/user/UserResourceTest.java`.

Let us now analyze the most relevant test cases.

Testing the user list endpoint

This test verifies that a user with an administrative role can retrieve a list of all the users. The following snippet contains the test source code:

```
@Test
@TestSecurity(user = "admin", roles = "admin")
void list() {
  given()
    .when().get("/api/v1/users")
    .then()
```

```
        .statusCode(200)
        .body("$.size()", greaterThanOrEqualTo(1),
          "[0].name", is("admin"),
          "[0].password", nullValue());
}
```

If you recall, in *Chapter 3*, *Creating the HTTP API*, we annotated the UserResource class with the @RolesAllowed annotation and secured all of its declared endpoints so that only authenticated users with the admin role could perform requests. To test the user endpoints, we could retrieve a valid JWT and include it as an HTTP request header. However, Quarkus provides a very handy @TestSecurity annotation to control the application's security context when the test is run. The annotation can be used to completely disable authorization so that secured endpoints can be accessed without being authenticated or to specify the identity and roles of the application's context user for requests performed during the annotated test execution.

For the current test, we want to verify that a user with an admin role can retrieve the list of users, so we'll annotate the method with @TestSecurity(user = "admin", roles = "admin").

The test implementation performs a simple HTTP GET request to the /api/v1/users endpoint and verifies that a 200 OK status code is returned. The test also verifies that the JSON response contains at least a user with the name admin and that the user's password is *not* leaked. Notice how REST Assured allows you to perform **JSONPath** queries to the response body and validate them with expression matchers. Let us now see how to verify the user creation features.

Testing the user creation features

These tests verify that only users with the admin role can create new users and assign them roles. The following snippet contains the source code for a test that verifies the **happy path** for the user creation procedure:

```
@Test
@TestSecurity(user = "admin", roles = "admin")
void create() {
  given()
    .body("{\"name\":\"test\",\"password\":\"test\",
      \"roles\":[\"user\"]}")
    .contentType(ContentType.JSON)
    .when().post("/api/v1/users")
    .then()
    .statusCode(201)
    .body(
```

```
        "name", is("test"),
        "password", nullValue(),
        "created", not(emptyString())
    );
}
```

The test method is annotated with the `@TestSecurity` annotation configured with a user with the `admin` role since only administrators can create users. The implementation starts by building an HTTP request with a JSON body and content type. The body contains the minimum required fields to create a valid user, in this case, `name`, `password`, and `roles`. The HTTP request is sent with a `POST` method to the user's resource endpoint. In the assertion phase, we validate that the HTTP response returns a `201 Created` successful status code. Then, we verify that the response contains a JSON body with the new user's information, including the automatically generated `created` field, and that the password is not leaked.

> **Happy path**
>
> The happy path, in the context of testing, is the scenario that showcases a feature's workflow when no exception occurs and the action completes successfully.

In the happy path scenario, we have verified that a user with the `admin` role can create a user. Let us now test that a regular user *can't* invoke the endpoint. The following code snippet contains the code that verifies this:

```
@Test
@TestSecurity(user = "user", roles = "user")
void createUnauthorized() {
    given()
        .body("{\"name\":\"test-unauthorized\",\"password\"
          :\"test\",\"roles\":[\"user\"]}")
        .contentType(ContentType.JSON)
        .when().post("/api/v1/users")
        .then()
        .statusCode(403);
}
```

In this case, the test method is also annotated with the `@TestSecurity` annotation, but the configured user doesn't have the `admin` role. The test method implementation starts just like the happy path alternative. However, in the assertion phase, we check that the HTTP response contains a `403 Forbidden` client error status code.

Since the `User` entity has a unique constraint for the `name` field, we can also create a scenario to verify that the constraint is enforced. The following snippet contains the source code for a test to verify it:

```
@Test
@TestSecurity(user = "admin", roles = "admin")
void createDuplicate() {
    given()
        .body("{\"name\":\"user\",\"password\":
          \"test\",\"roles\":[\"user\"]}")
        .contentType(ContentType.JSON)
        .when().post("/api/v1/users")
        .then()
        .statusCode(409);
}
```

In this case, both the method annotations and the HTTP request data are configured correctly. However, the requested user name is a duplicate of one of the users we create in the `import-dev.sql` file. In the verification phase, we assert that the response has a `409 Conflict` client error status code. In this case, we're not only verifying that the user-related features work OK, but that the `RestExceptionHandler` class we implemented in *Chapter 3, Creating the HTTP API*, is processing the exception correctly.

We have now tested all the scenarios for the user creation feature. Let us continue by implementing tests for the user update functionality.

Testing the user update feature

These tests verify that users with the `admin` role can update existing users. The following test asserts that a recently created user can be modified to change its name to a different value. The following snippet contains the source code for this test:

```
@Test
@TestSecurity(user = "admin", roles = "admin")
void update() {
    var user = given()
        .body("{\"name\":\"to-update\",\"password\":
          \"test\",\"roles\":[\"user\"]}")
        .contentType(ContentType.JSON)
        .when().post("/api/v1/users")
        .as(User.class);
```

```
    user.name = "updated";
    given()
       .body(user)
       .contentType(ContentType.JSON)
       .when().put("/api/v1/users/" + user.id)
       .then()
       .statusCode(200)
       .body(
          "name", is("updated"),
          "version", is(user.version + 1)
       );
}
```

Only users with the admin role can interact with this endpoint, so we annotate the test method with a properly configured @TestSecurity annotation. The test implementation differs a little bit from the ones we've analyzed so far. You'll notice that it contains two REST Assured invocations, one to prepare the test scenario and another one to perform the actual test and its assertions.

The first REST Assured invocation is used to create a new user with the name to-update. The code is very similar to the one we implemented when developing the user creation feature tests. However, in this case, instead of providing a then() DSL method assertion, the HTTP response body is deserialized into a User instance.

The next part contains the actual test, where the user with an updated name field value is sent to the application with a PUT HTTP request. The assertion part verifies that the server responds with a 200 OK status code and that the response body contains the updated user information with an incremented value in the version field.

In the *Creating the task manager entities* in *Chapter 2*, *Adding Persistence*, we learned that the version field was used by Hibernate to provide the optimistic locking features. We can also implement a test to verify that optimistic locking works and concurrent entity modifications are prevented by our application. The next code snippet contains the source code for this test:

```
@Test
@TestSecurity(user = "admin", roles = "admin")
void updateOptimisticLock() {
    given()
       .body("{\"name\":\"updated\",\"version\":1337}")
       .contentType(ContentType.JSON)
       .when().put("/api/v1/users/0")
       .then()
```

```
        .statusCode(409);
    }
```

The test implementation performs an HTTP request to update the value of the user with the 0 ID (admin). However, we are providing a fake version number that doesn't match the one in the database. The assertion checks that the server responds with a `409 Conflict` client error status code.

There's an additional test case for the user update operation when the user doesn't exist, which you can check out in the book's source code repository.

Let us now analyze the test for the delete operation.

Testing the user delete feature

This test verifies that a user with an administrative role can delete an existing user from the database. In the following snippet, you can find the source code for this test:

```
@Test
@TestSecurity(user = "admin", roles = "admin")
void delete() {
    var toDelete = given()
        .body("{\"name\":\"to-delete\",\"password\":\"test\"}")
        .contentType(ContentType.JSON)
        .post("/api/v1/users")
        .as(User.class);
    given()
        .when().delete("/api/v1/users/" + toDelete.id)
        .then()
        .statusCode(204);
    assertThat(User.findById
        (toDelete.id).await().indefinitely(), nullValue());
}
```

The test method is configured with the `@TestSecurity` annotation, so the test is executed in the context of a logged-in user with the admin role. Just like we did for the update scenarios in the *Testing the user update feature* section, the test implementation starts by creating a user using a REST Assured configured HTTP operation that we'll subsequently delete. The next invocation is the actual test where we send a DELETE HTTP request and verify that the server responds with a `204 No Content` successful status code. The test implementation contains an additional assertion that verifies that the user no longer exists in the database.

We have now implemented tests for most of the CRUD operation scenarios of the user resource. Let us now prepare a test for the password change operation.

Testing the password change feature

This test verifies that any application user can change their password using the provided endpoint. The following snippet contains the source code for this test:

```
@Test
@TestSecurity(user = "admin", roles = "user")
void changePassword() {
  given()
    .body("{\"currentPassword\": \"quarkus\",
      \"newPassword\": \"changed\"}")
    .contentType(ContentType.JSON)
    .when().put("/api/v1/users/self/password")
    .then()
    .statusCode(200);
  assertTrue(BcryptUtil.matches("changed",
    User.<User>findById(0L).await().indefinitely().password)
  );
}
```

When we implemented the `UserResource` class in *Chapter 3, Creating the HTTP API*, we declared the `changePassword` method with an `@RolesAllowed("user")` annotation, which allows any user with the `user` role to invoke this endpoint. For the test, we'll annotate the test case method with an `@TestSecurity` annotation with the `roles` property set to `user`. The test implementation prepares an HTTP request with a JSON body that contains two fields: `currentPassword` and `newPassword`. Then, we send the request using an HTTP `PUT` method and verify that the server responds with a `200 OK` successful status code. We'll provide an additional assertion to check that the user's password was changed in the database and set to the corresponding bcrypt encrypted value.

The test suite contains an additional test case, `changePasswordDoesntMatch()`, which contains a similar implementation but with an invalid `currentPassword` field value. The assertion checks that the invocation fails with a `409 Conflict` client error HTTP status code.

Now that we've covered most of the user-related test cases, let us continue by implementing the tests for the project-related features.

Testing the project-related features

In the *ProjectService* and *ProjectResource* sections in *Chapter 3, Creating the HTTP API*, we developed and exposed the project-related features. We added endpoints to list, create, update, and delete projects for the currently logged-in user. We will now implement tests to verify that these features work as expected. Just like we did for the `user` and `auth` resource classes in the *Testing the user-related features* and *Testing authentication* sections, we'll open the `ProjectResource` class, right-click on its declaration, click on the **Go To** submenu, and then the **Test** menu entry. We'll click on the **Create New Test** popup, and accept the default dialog settings. IntelliJ should now create and open a new test `ProjectResourceTest` class.

You can find the full source code for the test suite at `https://github.com/PacktPublishing/Full-Stack-Development-with-Quarkus-and-React/tree/main/chapter-05/src/test/java/com/example/fullstack/project/ProjectResourceTest.java`. Most of the test cases are very similar to those that we implemented for the user-related features. However, the project delete operation is different from that of the user's homologous delete operation; let us analyze this test case.

Testing the project delete feature

This test verifies that a user can delete one of their projects and that any task assigned to the project is reassigned so that it no longer references the deleted project. The following code snippet contains the source code for this test case:

```java
@Test
@TestSecurity(user = "user", roles = "user")
void delete() {
  var toDelete = given().body("{\"name\":\"to-delete\"}").
    contentType(ContentType.JSON)
    .post("/api/v1/projects").as(Project.class);
  var dependentTask = given()
    .body("{\"title\":\"dependent-task\",\"project\":
      {\"id\":" + toDelete.id + "}}").contentType
        (ContentType.JSON)
    .post("/api/v1/tasks").as(Task.class);
  given()
    .when().delete("/api/v1/projects/" + toDelete.id)
    .then()
    .statusCode(204);
  assertThat(Task.<Task>findById(dependentTask.id).await().
```

```
        indefinitely().project, nullValue());
    }
```

The test method is annotated with the `@TestSecurity` annotation with the `user` and `roles` properties configured. For these tests, the `user` property must point to an existent user, since all of the project operations are performed within the scope of an owning user. The test implementation starts by creating a project and a dependent task using real HTTP requests performed with REST Assured. The third REST Assured instruction contains the actual test, where we send a `DELETE` HTTP request to the project's resource endpoint: `"/api/v1/projects/" + toDelete.id`. We then verify that the server responds with a `204 No Content` successful status code. We perform an additional assertion to verify that the dependent task stored in the database no longer references the deleted project.

We've now covered the test suites for the authentication, user, and project-related features. Let us now implement the tests for the task-related functionality.

Testing the task-related features

In the *TaskService* and *TaskResource* sections in *Chapter 3, Creating the HTTP API*, we developed and exposed the task-related features. We added endpoints to create, update, and delete tasks for the currently logged-in user, and also a specific endpoint to set the completion status of a task. We will now implement tests to verify that these features work as expected.

To create the test suite class, we'll open the `TaskResource` class, right-click on its declaration, and click on the **Go To** submenu, and the **Test** menu entry. We'll click on the **Create New Test** popup, and accept the default dialog settings. IntelliJ should now create and open a new `TaskResourceTest` class.

You can find the full source code for the test suite at `https://github.com/PacktPublishing/Full-Stack-Quarkus-and-React/blob/main/chapter-05/src/test/java/com/example/fullstack/task/TaskResourceTest.java`. Most of the test cases are very similar to those that we implemented for the other resources. However, the set task complete operation has a subtle difference, since it expects a subresource URL and a `boolean` body instead of a complete object. Let us analyze the test to verify the set task complete operation.

Testing the set task complete feature

This test verifies that a user can mark one of their tasks as complete. The following code snippet contains the source code for this test case:

```
@Test
@TestSecurity(user = "user", roles = "user")
void setComplete() {
    var toSetComplete = given()
```

```
    .body("{\"title\":\"to-set-complete\"}")
    .contentType(ContentType.JSON)
    .post("/api/v1/tasks").as(Task.class);
given()
    .body("\"true\"")
    .contentType(ContentType.JSON)
    .when().put("/api/v1/tasks/" + toSetComplete.id + "/
complete")
    .then()
    .statusCode(200);
assertThat(Task.findById(toSetComplete.id).await().
    indefinitely(),
    allOf(
      hasProperty("complete", notNullValue()),
      hasProperty("version", is(toSetComplete.version + 1))
    ));
}
```

We start the test implementation by creating a task using REST Assured to perform an HTTP POST request. Next, we send a PUT request to the task complete resource endpoint (`"/api/v1/tasks/"` `+ toSetComplete.id + "/complete"`) and verify that the server responds with a `200 OK` successful HTTP status code. We include an additional assertion to verify that the task in the database is updated and the `complete` field is set to a non-null value.

Now that we've provided full coverage for our backend, we can execute the tests using Maven by running the following command:

```
./mvnw clean test
```

Maven should execute your tests and you should be able to see the following message:

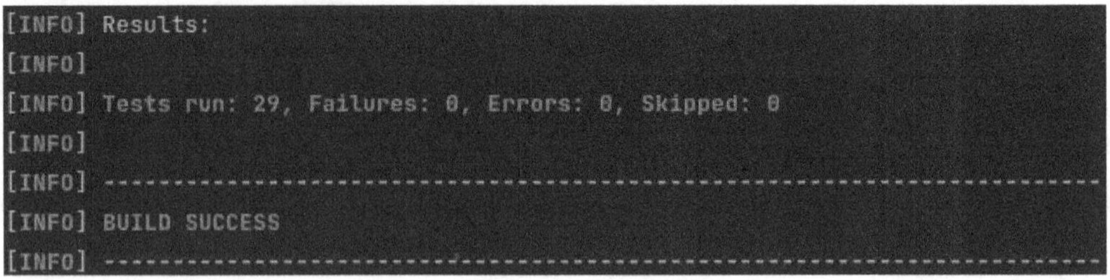

```
[INFO] Results:
[INFO]
[INFO] Tests run: 29, Failures: 0, Errors: 0, Skipped: 0
[INFO]
[INFO] ------------------------------------------------------------
[INFO] BUILD SUCCESS
[INFO] ------------------------------------------------------------
```

Figure 5.8 – A screenshot of the execution result of the mvnw clean test command

We have now implemented integration tests for all of the application's resources covering most of the usage scenarios. Our application is now much more reliable, and we should feel safe when we need to refactor or maintain the project in the future.

Summary

In this chapter, we learned how to implement tests in Quarkus using its test framework and REST Assured. We also provided complete test coverage for the task manager backend. We started by adding the missing dependency and configuring the project. Then, we added test suite classes for the auth, user, project, and task resources. We also learned how to run these tests using IntelliJ's built-in features, and through Maven in the command line.

You should now be able to implement tests for your Quarkus applications using the Quarkus test framework and REST Assured. In the next chapter, we'll learn how to compile the application to a native executable using GraalVM.

Questions

1. What configuration property should you use to add data to your testing environment database?
2. How do you bootstrap a test suite class in IntelliJ?
3. How does an `@QuarkusTest` differ from a regular JUnit test?
4. What's the purpose of the `@TestSecurity` annotation?

6

Building a Native Image

In this chapter, we'll study how to compile a native executable file for the task manager. We'll start by learning the main advantages and benefits of Quarkus, GraalVM, and their native image compilation features. Then, we'll learn how to set up GraalVM in our system. Next, we'll configure our application for native compilation, build it, and run the resulting native executable. Finally, we'll learn how to produce a Linux-compatible executable without having a local GraalVM installation.

By the end of this chapter, you should be able to compile your Quarkus applications into native executable images and have a good understanding of why and when you should choose this packaging method.

> **GraalVM Native Image**
>
> GraalVM is a high-performance **Java Development Kit** (**JDK**) and **Java Virtual Machine** (**JVM**) originally developed and maintained by Oracle. Some of its most notable features among many are a high-performance compiler, **Ahead-Of-Time** (**AOT**) native image compilation, and polyglot programming. GraalVM Native Image is an AOT compilation technology that generates standalone native executables. The executable packages all the necessary classes from your application and their dependencies, along with statically linked native code from the JDK. This means that it's self-contained and can be executed without any additional requirements.

We will be covering the following topics in this chapter:

- Building a native executable in Quarkus
- Creating a native image for the task manager

Technical requirements

You will need the latest Java JDK LTS version (at the time of writing, Java 17). In this book, we will be using Fedora Linux, but you can use Windows or macOS as well.

You will need a working Docker environment to deploy a PostgreSQL database and to create a Linux native image using Docker. There are Docker packages available for most Linux distributions. If you are on a Windows or macOS machine, you can install Docker Desktop.

In case you're not using IntelliJ IDEA Ultimate edition, you'll need a tool such as **cURL** or **Postman** to interact with the implemented HTTP endpoints.

You can download the full source code for this chapter from `https://github.com/ PacktPublishing/Full-Stack-Quarkus-and-React/tree/main/chapter-06`.

Building a native executable in Quarkus

In the *What is Quarkus?* section in *Chapter 1, Bootstrapping the Project*, we learned that Quarkus was built from the ground up to be cloud-native by considerably improving the application's *startup time* and *memory footprint*. To improve the application's boot time, Quarkus moves many of the tasks that classic Java applications perform at runtime to build time.

Every time a *traditional* Java application boots, it loads and analyzes configuration files, scans for annotations, builds a dependency tree, and so on, before even starting with the application's execution logic. Quarkus, on the other hand, rethinks the problem and moves most of these tasks to build time where the results are recorded as part of the application's bytecode. This way, Quarkus can start with the application's execution logic right away from the moment the application boots, which considerably improves the overall startup time and memory consumption.

With this approach, Quarkus brings substantial performance improvements to Java runtimes executed in a traditional JVM. However, even faster startup times and smaller memory footprints can be achieved by using AOT compilation and creating a native executable. Quarkus leverages GraalVM Native Image AOT compilation to compile and package native binaries.

Combining GraalVM's AOT compiler optimizations with the Quarkus build time approach delivers application runtimes with the smallest possible memory footprint and startup time. Low memory footprint applications are especially interesting when deploying them in cloud environments such as Amazon or Google since it will drastically reduce the costs associated with their deployment. Fast startup times are crucial for cloud-native applications that need to scale very quickly or that are delivered as serverless functions.

> **Note on performance**
>
> Packaging your application as a native executable brings the added benefits of faster boot time and reduced memory footprint. However, this comes with a trade-off concerning the overall application performance. When a Java application is run on a JVM, it takes advantage of runtime optimizations that the virtual machine performs by building profile information throughout its execution lifetime. When choosing a packaging strategy for your application, you need to compare and balance the benefits of each approach and choose the one that better suits your use case. For example, you'd usually prefer native executable packaging for smaller, short-lived applications.

GraalVM can be used to compile any Java application. However, it requires fine-tuning and tedious configuration to make it work. Even if you follow the instructions and provide careful configuration, you might still not be able to create a native image of your application. Following its developers' joy paradigm we learned about in the *What is Quarkus?* section, in *Chapter 1, Bootstrapping the Project*, Quarkus and its extensions provide many of the necessary steps and configurations for native image compilation, so that developers can concentrate on the code and not on the build process. With Quarkus, you'll still likely need to set some configuration options or annotate your code, but it's nothing compared to what you'll require with other frameworks or vanilla Java code.

Now that we've learned the Quarkus native image compilation advantages, let us see how to set up GraalVM in our systems.

Setting up GraalVM

Quarkus takes advantage of GraalVM to compile your applications into native image executables. To get started, you first need a valid GraalVM setup on your system. There are three distributions of GraalVM:

- Oracle GraalVM **Community Edition (CE)**

- Oracle GraalVM **Enterprise Edition (EE)**

- Mandrel: A GraalVM distribution, specifically designed to provide native image support for Quarkus

In this section, we'll learn how to manually set up GraalVM CE in a Linux environment step by step:

1. Download a Java 17 GraalVM CE package for your environment from the GitHub releases page: `https://github.com/graalvm/graalvm-ce-builds/releases`. In my case, it's Linux (amd64).

2. Extract the archive to its definitive location. I will be extracting it to the `$HOME/lib/graalvm-ce` directory.

3. Set the `GRAALVM_HOME` environment. You can achieve this by typing the following command:

```
export GRAALVM_HOME=$HOME/lib/graalvm-ce
```

4. Add the GraalVM `bin` subdirectory to your path:

```
export PATH=${GRAALVM_HOME}/bin:$PATH
```

Note that you should save these commands in your `.bashrc` file to make them persistent.

5. Install GraalVM's `native-image` tool:

```
gu install native-image
```

The `native-image` tool should be automatically downloaded and installed, and you should see a message like the following:

```
$ gu install native-image
Downloading: Component catalog from www.graalvm.org
Processing Component: Native Image
Downloading: Component native-image: Native Image from github.com
Installing new component: Native Image (org.graalvm.native-image, version 22.1.0)
```

Figure 6.1 – A screenshot of the result of executing gu install native-image

We should now have a valid GraalVM CE setup on our Linux-compatible system. Let's verify this by running the `java -version` command:

```
$ java -version
openjdk version "17.0.3" 2022-04-19
OpenJDK Runtime Environment GraalVM CE 22.1.0 (build 17.0.3+7-jvmci-22.1-b06)
OpenJDK 64-Bit Server VM GraalVM CE 22.1.0 (build 17.0.3+7-jvmci-22.1-b06, mixed mode, sharing)
```

Figure 6.2 – A screenshot showing the GraalVM CE build of OpenJDK as the current Java version

You should now be able to set up GraalVM on Linux by yourself. The `graalvm.org` website also contains detailed documentation on how to install it on Windows and macOS platforms. In addition, there are also platform-specific packages (`sdkman`, `homebrew`, `scoop`, and so on) you can use to automatically install GraalVM on your system. Regardless of how you install GraalVM on your system, note that you will still need to install GraalVM's `native-image` tool manually, as stated in *step 5*: `gu install native-image`.

A local GraalVM setup is required to build a native image for *your platform*. However, Quarkus also supports building a Linux-compatible native image in a Docker container without the need for a local GraalVM installation. In the *Building the native image in a docker container* section, we'll use this technique to create a Linux executable for our task manager application.

We now have a good understanding of the advantages of compiling our Java application into a native image and a GraalVM setup. Let us continue by creating a native image for the task manager we are implementing across the book.

Creating a native image for the task manager

In general, building a native executable with GraalVM usually requires tedious configuration and tweaking. Instead, when building an executable for a Quarkus application, Quarkus does most of the heavy lifting for us. However, we'll still need to tweak our application a little.

Let us start by configuring the application to account for additional application resources that might be missed by GraalVM.

Including application resources

GraalVM Native Image uses AOT compilation to generate the native executable. This means that GraalVM shifts most of the processes and analyzes that regular Java applications perform during the runtime/execution phase to the build/compilation phase. This is the main reason why GraalVM compiled native applications perform so much better than regular Java applications. To accomplish this, GraalVM needs to know at build time everything that will be required at runtime so that it gets included in the binary executable. The Native Image builder performs an aggressive static code analysis to find all of the methods that are reachable from the application's entry point. Then, it compiles *only* these methods into the application's executable.

Considering this, GraalVM will require extra help when dealing with dynamic class loading, reflection, serialization, and so on. As we learned in the *Building a native executable in Quarkus* section, Quarkus and its extensions take care of providing most of these settings. For the task manager, most of it is taken care of except for the **JSON Web Token** (**JWT**) keys we created in *Chapter 4*, *Securing the Application*. We need to add the following entry to the application.properties file to take care of these files:

```
quarkus.native.resources.includes=jwt/public-key.pem,
jwt/private-key.pem
```

The quarkus.native.resources.includes property can be configured with a comma-separated list of paths that will be added to the native image. In this case, we're including the public key that is used by the task manager to verify the JWTs along with the private key that is used by the task manager to sign them. We'll learn an additional method to include resources in the native image when we integrate the frontend with Quarkus in *Chapter 11*, *Quarkus Integration*. Our application should now be ready for packaging, so let us build the native image.

Building the native image

In the *Maven project (pom.xml)* section in *Chapter 1*, *Bootstrapping the Project*, we learned that the https://code.quarkus.io site bootstrapped a project with a configured native Maven profile for us. This profile contains all the required settings to package the application as a native image using a local GraalVM installation. We don't need to provide further changes to this profile, so we can start building the native image right away.

The process to build a native image is very similar to the one used to package the standard JVM artifact. We'll need to invoke the Maven `package` goal but with the native profile instead. Let's try to compile the application by running the following command:

```
./mvnw clean package -Pnative
```

The process should complete successfully and we should be able to see a build result summary similar to the following:

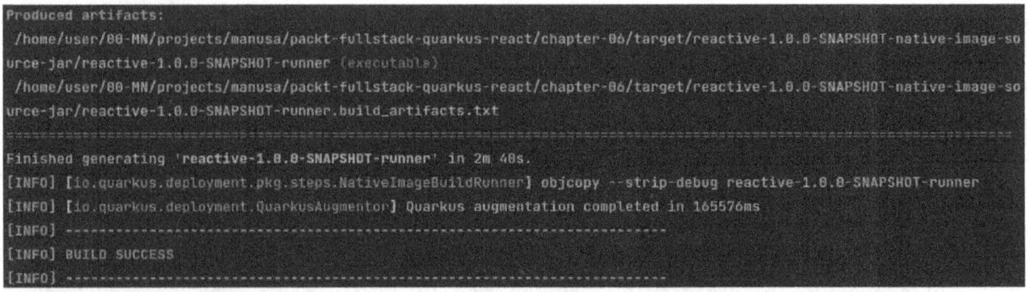

```
Produced artifacts:
 /home/user/00-MN/projects/manusa/packt-fullstack-quarkus-react/chapter-06/target/reactive-1.0.0-SNAPSHOT-native-image-so
urce-jar/reactive-1.0.0-SNAPSHOT-runner (executable)
 /home/user/00-MN/projects/manusa/packt-fullstack-quarkus-react/chapter-06/target/reactive-1.0.0-SNAPSHOT-native-image-so
urce-jar/reactive-1.0.0-SNAPSHOT-runner.build_artifacts.txt
===============================================================================
Finished generating 'reactive-1.0.0-SNAPSHOT-runner' in 2m 40s.
[INFO] [io.quarkus.deployment.pkg.steps.NativeImageBuildRunner] objcopy --strip-debug reactive-1.0.0-SNAPSHOT-runner
[INFO] [io.quarkus.deployment.QuarkusAugmentor] Quarkus augmentation completed in 165576ms
[INFO] ------------------------------------------------------------------------
[INFO] BUILD SUCCESS
[INFO] ------------------------------------------------------------------------
```

Figure 6.3 – A screenshot of the result of executing ./mvnw clean package –Pnative

A new executable file, `reactive-1.0.0-SNAPSHOT-runner`, with the compiled application should have been created in the `target` directory. Let us now see how to execute this file and verify that the application works.

Running the native image

So far we've been executing the application in *development* mode and have been taking advantage of Quarkus dev services to create and configure a PostgreSQL database for us. However, the native image executable contains the *production* code, so we need to have a database available and provide its configuration parameters to the application runtime.

If you don't have a PostgreSQL instance in your local environment, you can create one in a Docker container. Let's try this by running the following command:

```
docker run --rm --name postgresql -p 5432:5432 -e POSTGRES_
PASSWORD=pgpass -d postgres
```

This command starts a Docker container named `postgresql` and maps the container `5432` port (PostgreSQL default port) to the local `5432` port. The `--rm` flag instructs Docker to remove this container as soon as we stop it. Once we complete the test, you can stop and automatically remove the container by running `docker stop postgresql`.

We can now execute the application by running the following command:

```
./target/reactive-1.0.0-SNAPSHOT-runner -Dquarkus.datasource.
username=postgres -Dquarkus.datasource.password=pgpass
-Dquarkus.datasource.reactive.url=postgresql://localhost:5432/
postgres -Dquarkus.hibernate-orm.database.generation=create
```

This command starts the application and provides the required configuration options for the data source. The credentials and the URL should match those of your PostgreSQL database. In this case, we're providing configuration options that match the ones of the `postgresql` container we just created. In addition, we're instructing Hibernate to create the database for us in case it doesn't exist by overriding the `quarkus.hibernate-orm.database.generation` property with a `create` value. The application should start and we should be able to see the following log message:

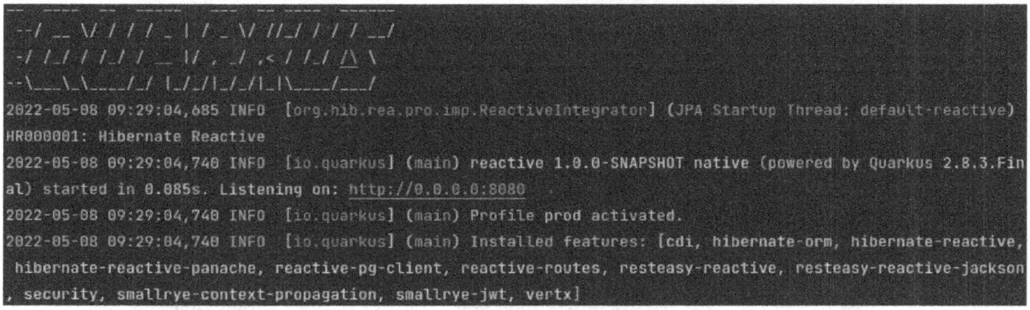

Figure 6.4 – A screenshot of the native application execution log

We can now test whether everything works OK by requesting a JWT for the `admin` user. Let's try this by running the following command:

```
curl -X POST -d"{\"name\":\"admin\",\"password\":\"quarkus\"}"
-H "Content-Type: application/json" localhost:8080/api/v1/auth/
login
```

If everything goes well, a new token should be printed:

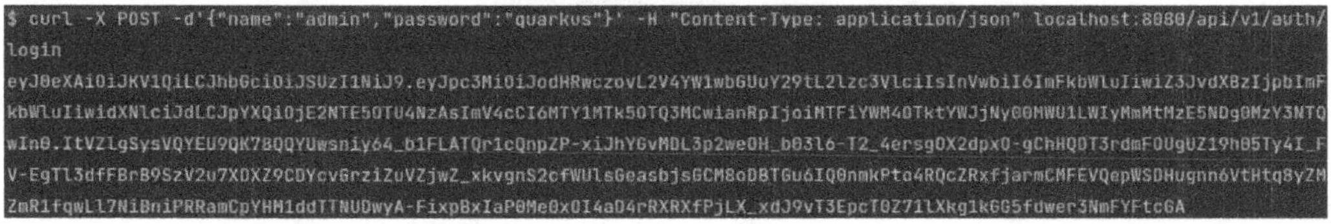

Figure 6.5 – A screenshot of the result of executing cURL to perform a valid login

> **Note**
>
> You need to manually stop the PostgreSQL Docker container we've created to test run the application (`docker stop postgresql`) when you're finished. If this container is active while running or testing the application in dev mode, you'll face port conflict errors.

We've now covered how to configure, compile, and run the task manager as a native executable. Let us now learn how to build a Linux executable from any platform using a Docker container.

Building the native image in a Docker container

In general, the main purpose of a native executable is to distribute it within a container image that can then be deployed into the cloud. For this scenario, you'll need a Linux-compatible binary file that might not match your local development platform. In this case, Quarkus doesn't even need you to have a valid GraalVM installation in your system, since it can perform the build leveraging a Docker container.

We can instruct Quarkus to perform a native image container build by setting the `quarkus.native.container-build` property to `true`. We should be able to do this in the command line when executing the Maven command like in the following example:

```
./mvnw clean package -Pnative -Dquarkus.native.container-build=true
```

However, in our case, whenever we build the native image, we want to create a Linux executable, so we'll make this change *permanent* by adding the property to the `native` profile in our project's `pom.xml`.

> **Note**
>
> At the time of writing, the default builder image Quarkus uses is based on Java 11, which is not compatible with our Java 17-based project. We'll need to configure a valid builder image in `pom.xml` by setting the `quarkus.native.builder-image` property too.

The following code snippet shows the resulting `native` profile properties section with the persisted property in `pom.xml` (you can find the full source code for this file at `https://github.com/PacktPublishing/Full-Stack-Development-with-Quarkus-and-React/tree/main/chapter-06/pom.xml`):

```
<profile>
  <id>native</id>
  <activation>
    <property>
      <name>native</name>
    </property>
```

```
  </activation>
  <build>
    <!-- … -->
  </build>
  <properties>
    <quarkus.package.type>native</quarkus.package.type>
    <quarkus.native.container-build>true
      </quarkus.native.container-build>
    <quarkus.native.builder-image>quay.io/quarkus/ubi-
      quarkus-mandrel:21.3-java17</quarkus.native.
        builder-image>
  </properties>
</profile>
```

We've added the `quarkus.native.container-build` property to the `native` profile properties section. Whenever we run Maven with this `native` profile enabled, the property will be applied automatically, and we won't need to provide it manually anymore.

Let's try to compile our project again by executing the following Maven command:

```
./mvnw clean package -Pnative
```

The process should complete successfully and we should be able to see a build result summary similar to the following:

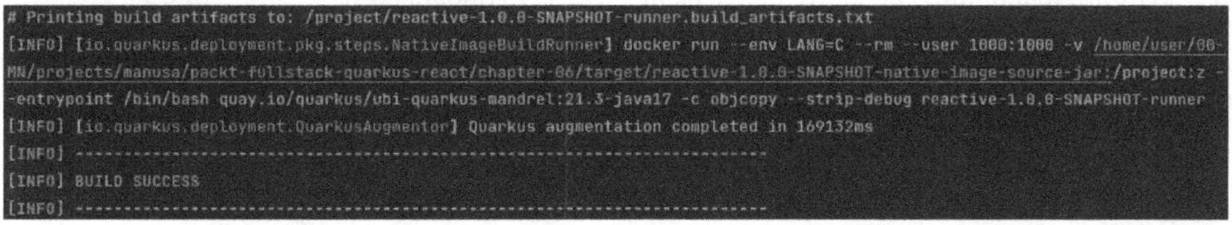

```
# Printing build artifacts to: /project/reactive-1.0.0-SNAPSHOT-runner.build_artifacts.txt
[INFO] [io.quarkus.deployment.pkg.steps.NativeImageBuildRunner] docker run --env LANG=C --rm --user 1000:1000 -v /home/user/86-
MN/projects/manusa/packt-fullstack-quarkus-react/chapter-06/target/reactive-1.0.0-SNAPSHOT-native-image-source-jar:/project:z -
-entrypoint /bin/bash quay.io/quarkus/ubi-quarkus-mandrel:21.3-java17 -c objcopy --strip-debug reactive-1.0.0-SNAPSHOT-runner
[INFO] [io.quarkus.deployment.QuarkusAugmentor] Quarkus augmentation completed in 169132ms
[INFO] -----------------------------------------------------------
[INFO] BUILD SUCCESS
[INFO] -----------------------------------------------------------
```

Figure 6.6 – A screenshot of the result of executing ./mvnw clean package –Pnative using a container build

Notice how the log shows two Docker invocations, one to perform the build and another to copy the resulting artifact from the Docker volume to the target directory. Regardless of your local development platform, a new `reactive-1.0.0-SNAPSHOT-runner` binary file should be available in the `target` directory.

Summary

In this chapter, we studied how to create a native image executable in Quarkus using GraalVM. We learned about the advantages of AOT compilation and native images. Then we learned how to set up GraalVM on a Linux platform and discovered other alternatives to set it up on Windows and macOS. Next, we configured, built, and ran our task manager application using native compilation and our local GraalVM installation. Finally, we analyzed how to perform the native image build using a Docker container and when you should choose this approach.

You should now be able to compile your Quarkus applications into native image executables by using a local GraalVM setup or a Docker container.

In the next chapter, we'll start implementing the frontend of the task manager. We'll bootstrap the project and give a basic introduction to React and the tools and libraries we'll be using in the next part of the book.

Questions

1. What is GraalVM Native Image?
2. Why do Quarkus applications have shorter startup times?
3. Does native image packaging always improve the application's performance?
4. What kind of applications benefit more from native image packaging?
5. Which Quarkus property can we use to include additional resources in the application's native image?
6. Why are we building the native image using a Docker container?

Part 2– Creating a Frontend with React

This section focuses on the knowledge and skills required to implement a frontend with React. In this part, you will learn how to create a React project from scratch to build the application's user interface and how to integrate it with the Quarkus backend.

In this part, we cover the following chapters:

- *Chapter 7, Bootstrapping the React Project*
- *Chapter 8, Creating the Login Page*
- *Chapter 9, Creating the Main Application*
- *Chapter 10, Testing Your Frontend*
- *Chapter 11, Quarkus Integration*

7

Bootstrapping the React Project

In this chapter, we'll bootstrap the frontend application for our project. We'll start by learning about React and its main advantages and differentiation points compared to alternative frameworks. Then, we'll learn how to create a React project from scratch and the recommended dependencies to start building a complete frontend application. Next, we'll create a common layout for the task manager that will be shared among the task manager pages to create a consistent look and feel. Finally, we'll create a temporary dummy page to check that all of the application's scaffolding features work as expected.

By the end of this chapter, you should have a basic understanding of React, frontend routing, global state management, and the advantages of using a styled component library. You should also be able to bootstrap an application from scratch that leverages several libraries to offer all of these features.

In this chapter, we will cover the following topics:

- What is React?
- Creating a React project
- Creating the common layout
- Displaying a dummy page

Technical requirements

You will need the latest **Node.js** LTS version (at the time of writing, 16.15). In this book, we will be using Fedora Linux, but you can use Windows or macOS as well.

You can download the full source code for this chapter from `https://github.com/PacktPublishing/Full-Stack-Quarkus-and-React/tree/main/chapter-07`.

What is React?

React is one of the most popular and widely adopted *open source* JavaScript libraries for building *user interfaces* with components. It was initially released by Facebook, now Meta, in 2013. Nowadays, it's still primarily maintained by Meta, although it has become a community project due to its extensive adoption.

React is a *declarative* library, which makes it simple and easy to use and provides a pleasant developer experience. Developers declare their **user interfaces** (**UIs**) and views for each state of their application, and React takes care of efficiently rendering and updating the appropriate components when the input data changes.

In contrast to other frontend frameworks such as Angular, React only focuses on component rendering and state management. To build a complete web application, developers will need to combine React with additional libraries to cover the rest of the features React doesn't provide. The main advantage is that developers have the freedom to select their stack, and have plenty of options to choose from. However, this also means that there are multiple solutions to the same problem. Depending on the technologies an application is based on, there might be more or less documentation and examples for such options.

Now that we have a good overview of React and its advantages, let's try to create and bootstrap a new project from scratch.

Creating a React project

There are many options and alternatives to creating a new React application project. In this book, we'll be using **Create React App**, since it is one of the easiest approaches both for the initial project creation and for its future maintenance. The main advantage of Create React App is that it provides all of the *build configuration and scripts* for you with a single project dependency. Under the hood, it uses popular tools such as **Webpack**, **Babel**, **ESLint**, and others. Another of its advantages is that there is *no lock-in*, so you can opt out at any point by performing an "eject" operation. When the project is "ejected," the single `react-scripts` dependency will be replaced by all of the required dependencies, and the build scripts will be generated and persisted in the source directory.

We can bootstrap the application by navigating to the `src/main` directory and executing the following command:

```
npx create-react-app frontend
```

The command should complete successfully, and you should be able to see the following log messages:

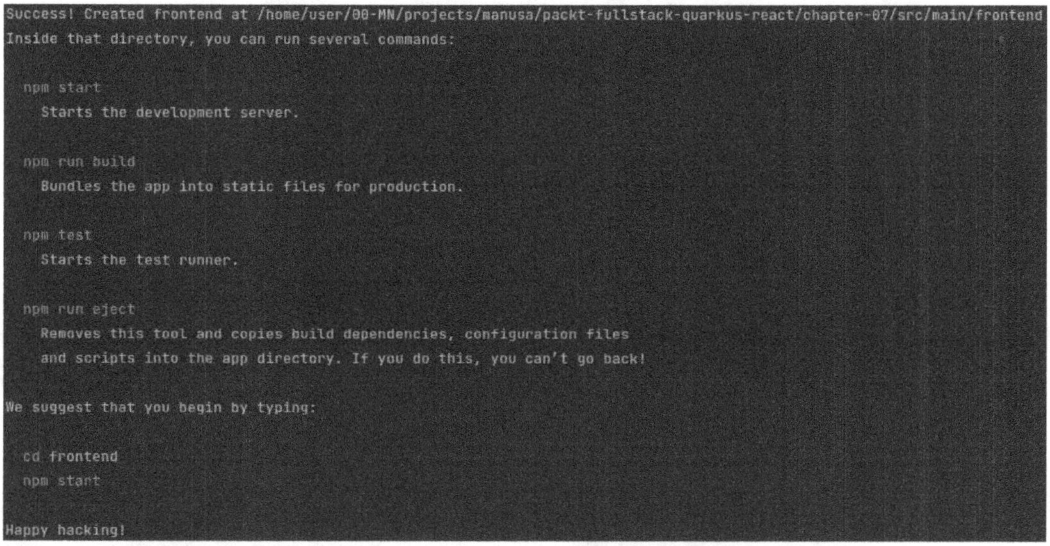

```
Success! Created frontend at /home/user/00-MN/projects/manusa/packt-fullstack-quarkus-react/chapter-07/src/main/frontend
Inside that directory, you can run several commands:

  npm start
    Starts the development server.

  npm run build
    Bundles the app into static files for production.

  npm test
    Starts the test runner.

  npm run eject
    Removes this tool and copies build dependencies, configuration files
    and scripts into the app directory. If you do this, you can't go back!

We suggest that you begin by typing:

  cd frontend
  npm start

Happy hacking!
```

Figure 7.1 – Screenshot of the result of executing npx create-react-app frontend

The new React application should be available under the `src/main/frontend` directory. We've chosen this location because we'll be distributing the application along with the Quarkus backend. In *Chapter 11, Quarkus Integration*, we'll integrate both projects and add the frontend build as part of the backend compilation process, and this location will make the integrating process much easier. Note that `src/main/frontend` is the root directory for the frontend application, so all of the future directory and file references in the rest of this chapter will be relative to this folder.

Create React App bootstraps a working and functional application. Let's try to start the application in development mode by navigating to the frontend application directory and executing the following command:

```
cd frontend
npm start
```

The command should start the application and we should be able to see the following message in the console:

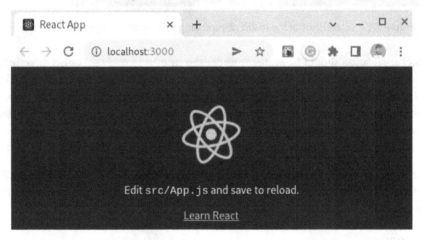

```
Starting the development server...
Compiled successfully!

You can now view frontend in the browser.

  Local:                http://localhost:3000
  On Your Network:      http://192.168.10.9:3000

Note that the development build is not optimized.
To create a production build, use npm run build.

webpack compiled successfully
```

Figure 7.2 – Screenshot of the result of executing npm start

A development server process should be up and running and we should be able to navigate to the application's home. Let's try to check the application in a browser by navigating to http://localhost:3000. If everything goes well, you should see the following page:

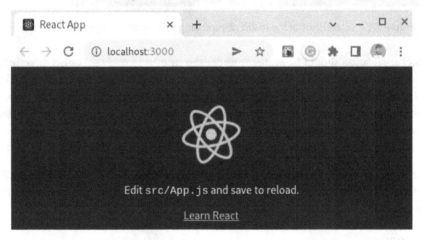

Figure 7.3 – A screenshot of a browser pointing to http://localhost:3000

Now, let's stop the server process by pressing *Ctrl* + *C* on our keyboard. Create React App configures the application with several scripts in the project's package.json file. We've already seen what happens with npm start, so let's go over what each of these commands does:

- npm start: This command invokes a script that runs the application in development mode. Any changes you make to the source code will force a page reload in the browser and will be reflected in the running application.

- `npm run build`: This command invokes the script that builds your application for production. It will transpile, bundle, and minimize your application and its dependencies. The final result will be available in the `build` directory, with the application ready to be deployed.

- `npm test`: This command invokes the test runner script. The script is configured to run the application's tests and stay in watch mode. This means that any change you make to the application code, or its tests, will automatically trigger a new test run.

- `npm run eject`: This command invokes the `eject` script. We've learned that Create React App has no lock-in. You can easily upgrade to a more complex build and script configurations by ejecting the application. This action will replace your project's `react-scripts` dependency with the individual dependencies for the required build tools. It will also persist all of the configurations and required files into your project. This action is permanent, which means that once you've ejected your application and tweaked your configuration files, you won't be able to move back to the Create React App approach. This is one of the greatest advantages of Create React App since you can bootstrap a project to get up and running quickly and move to complex builds once your application becomes more intricate.

Now that we've bootstrapped the frontend application and learned about the available scripts, let us analyze the project structure and the available files.

Directory structure

The Create React App script should have created a new project with a structure that looks like this:

Figure 7.4 – A screenshot of IntelliJ's UI showing the project's directory structure and files

At the project root, we can find the `package.json` file. This file contains the project configuration and its metadata. It is used to manage the project name, version, dependencies, scripts, and so on.

The `public` folder contains all of the assets that *won't* be processed by Webpack and will be directly copied during the build process to the `build` output directory. You can add assets that you want to distribute alongside the application without further processing to this directory. This folder also contains the `index.html` file, which is the page template and main entry point for the application. You can tweak this file to set the static page title, add additional meta tags, and so on.

The `src` directory contains the source code for the web application. Webpack will only process source files located under this subdirectory. This means that any `js` or `css` file needs to be in this directory to be taken into account for processing. When we implement the application, we'll add the source code files into the subdirectories of this folder to better organize the project.

The bootstrapped application source contains an `index.js` file, which is the main JavaScript entry point for the application. We will be adding logic and code that affects the *global* application to this file.

We can also find several `.css` files in the bootstrapped `src` directory. These files contain the styles for our application. In our case, for simplicity purposes, we'll be using **styled components** that won't require the use of additional styling (or, at most, very subtle changes).

There is also a complete sample component implementation, with its source code in the `App.js`, `App.css`, and `App.test.js` files:

- `App.js`: This file contains the component implementation – see *Figure 7.3* – which, at this point, is a spinning image with the message **Edit src/App.js and save to reload**. We will be modifying this component in the *Displaying a dummy page* section and using it as the routing entry point.

- `App.css`: This file contains the specific styles for this component. Since we'll be using styled-components, this file won't be necessary in the future.

- `App.test.js`: This file contains a simple test for the App component that verifies it renders and the message is displayed correctly. Note that the file is named with the `.test.js` suffix. Create React App testing scripts, which use **Jest** under the hood, search for files with this extension to run the tests they implement.

> **Jest**
>
> Jest is one of the most popular and widely adopted JavaScript testing frameworks. It allows you to write tests for both server backend applications and frontend applications. Jest provides full support for every aspect of testing: running the tests, mocking, setting expectations and assertions, and so on.

As we've learned, React only focuses on component rendering and its state management. We'll need to add some additional dependencies to provide routing, state management, and UI functionalities. Now, let's learn how to add frontend routing support by leveraging an additional library.

Adding routing

When implementing a frontend application, it's advisable to use a routing framework that allows you to define friendly URLs and helps deal with page navigation. Users of applications that have friendly routes can easily bookmark and share page locations. Neither React nor Create React App prescribe a specific routing solution. In our case, we'll be using **React Router**, which is one of the most popular options.

We'll start by adding the dependency to our project by executing the following command:

```
npm install react-router-dom@~6.3.0
```

The command should complete successfully, and the following line should be visible in the dependencies section of our package.json file:

```
"react-router-dom": "~6.3.0",
```

In the *Displaying a dummy page* section, we'll learn how to set up the routing for the application. Now, let's continue by adding a UI component library.

Adding the React Material UI component library

To speed up and simplify the task manager application development process, we'll be using a frontend component library. This will allow us to focus on developing the application functionalities without needing to deal with styling or implementing component UI interaction features. We will be using the **MUI Core** component library, which provides a rich set of components and tools based on Google's **Material Design** guidelines.

> **Material Design**
>
> Material Design is a system of guidelines and tools that supports the best practices and principles of good design when creating and designing user interfaces. It was created by Google in 2014 and was first introduced in Android 12. Its main purpose is to provide a consistent user experience for users of Google's products across all of its supported platforms, including the web and mobile.

Let's add the required dependencies to our project by executing the following command:

```
npm install @mui/material@~5.8.0 @emotion/react@~11.9.0 @
emotion/styled@~11.8.1 @mui/icons-material@~5.8.0
```

The command should complete successfully, and the following lines should be visible in the dependencies section of our `package.json` file:

```
"@emotion/react": "~11.9.0",
"@emotion/styled": "~11.8.1",
"@mui/icons-material": "~5.8.0",
"@mui/material": "~5.8.0",
```

Google's Material Design type system is based on the *Roboto* typography, which is not included as part of the MUI packages. The easiest way to include the font in the application is by using a markup link to load the font from Google's **content delivery network** (**CDN**). We can do this by editing the `public/index.html` file and including a link in the `head` section. The following code snippet shows the relevant parts:

```
<title>React App</title>
<link
    rel="stylesheet"
    href="https://fonts.googleapis.com/css?family=
        Roboto:300,400,500,700&display=swap"
/>
</head>
```

So far, we've created a React application and included additional libraries to provide routing and UI capabilities. Now, let's include a library to manage the global state of the application.

Adding state management

So far, we've learned that React can manage the state of the application's components. However, it doesn't provide a built-in, easy way to manage the *global* state of the application. A global state enables the application to share data between components easily. This allows components to both modify and react to changes performed on the global application state and the data it represents.

Redux is one of the most popular JavaScript libraries for centralizing and managing an application's global state. However, it involves a complex setup and requires verbose and repetitive boilerplate code. This has always raised criticism among its users and competitors and is the main reason why the maintainers created **Redux Toolkit**, the library we'll be using. Redux Toolkit is an opinionated distribution of Redux that includes utilities that simplify its configuration for the most common use cases.

Now, let's add the required dependencies to our project by executing the following command:

```
npm install @reduxjs/toolkit@~1.8.1 react-redux@~8.0.1
```

The command should complete successfully, and the following lines should be visible in the dependencies section of our `package.json` file:

```
"@reduxjs/toolkit": "~1.8.1",
"react-redux": "~8.0.1",
```

With that, we've added all of the required dependencies to our project and we're ready to start implementing the application's functionality.

Now, let's learn how to create a common layout for the application's pages.

Creating the common layout

In this section, we'll create a common layout component that will be used across the different application pages as a wrapper to provide a consistent look and feel. Our layout will include a top bar that contains the application's title, a home link, and an icon with a drop-down menu to perform user-specific actions. We'll also include a drawer with a menu containing entries to navigate to each of the application's pages.

We'll start by creating a `layout` directory beneath the `src` folder that will hold the components and code related to the layout. The application layout will consist of two basic components: a drawer and a top bar. Let's start by creating the `TopBar` component.

> **Note on the React component exports**
>
> When implementing the React components, we'll create specific JavaScript files for each of them. We'll also group the files and components into modules/directories based on the features they provide. To define a public API, easy imports, naming consistency, and some additional benefits, we'll export the components in an `index.js` file for each of the component directories. You can check the resulting directory structure and component modules in this book's GitHub repository: https://github.com/PacktPublishing/Full-Stack-Development-with-Quarkus-and-React/tree/main/chapter-07/src/main/frontend.

Creating the TopBar component

The `TopBar` component is used to display the current page title and has buttons to navigate to the home page, create a new task, and perform actions related to the logged-in user. The resulting component should look like this when rendered:

Figure 7.5 – A screenshot of the TopBar component

To implement the component, let's create a new file in the `src/layout` directory called `TopBar.js`, which will contain its source code. The following snippet contains the relevant part of the code (you can find the complete code in this book's GitHub repository at `https://github.com/PacktPublishing/Full-Stack-Quarkus-and-React/blob/main/chapter-07/src/main/frontend/src/layout/TopBar.js`):

```
export const TopBar = ({goHome, newTask, toggleDrawer}) => (
  <AppBar position='fixed' sx={{zIndex: theme =>
    theme.zIndex.drawer + 1}}>
    <Toolbar>
      <IconButton size='large' edge='start' color='inherit'
        aria-label='menu'
        onClick={toggleDrawer} >
        <MenuIcon />
      </IconButton>
      <Tooltip title='Home'>
        <IconButton color='inherit' onClick={goHome}>
          <HomeOutlinedIcon />
        </IconButton>
      </Tooltip>
      <Typography variant='h6' component='div' sx={{
        flexGrow: 1 }}>
        Task manager
      </Typography>
      <Tooltip title='Quick Add'>
        <IconButton color='inherit' onClick={newTask}>
          <AddIcon />
        </IconButton>
      </Tooltip>
      <AccountCircleIcon />
    </Toolbar>
  </AppBar>
);
```

In the first line, we're defining a *named export* (`export const TopBar`) and we are assigning the variable with a JavaScript arrow function that implements a React *functional component*. The component accepts three properties:

- `goHome`: Expects a function that, when invoked, will help the application navigate to the home page

- `newTask`: Expects a function that, when invoked, will start the new task creation workflow

- `toggleDrawer`: Expects a function that, when invoked, will collapse and uncollapse the application drawer alternately

The component implementation is based on the composition of several MUI components and icons using **JavaScript XML (JSX)** syntax. The main container is `AppBar`, which in Material Design is used to display information and actions related to the current screen. We define the `position` property with a `fixed` value; this means that whenever we scroll the page, the top bar will remain visible and won't scroll with the page. The `sx` property is common to all MUI components and is used to override or provide additional CSS styles. Since our layout combines a drawer with `AppBar`, we need to override the z-index to make sure that `AppBar` always appears on top of the drawer. Let's examine some of the other MUI components:

- `<Toolbar>`: This component is intended to be used along `AppBar` to act as a wrapper around the components it contains.

- `<IconButton>`: A type of button that contains a single icon and no text. They are commonly found in toolbars, or as toggle buttons.

- `<Tooltip>`: Tooltips are used to display informative messages whenever a user hovers over, focuses, or taps the element they surround.

- `<Typography>`: This is a convenient MUI component that can be used to present textual content to the user with different appearances and styles without needing to provide additional CSS configuration. In this case, we're using it to place the page title in `AppBar`.

JSX

JSX is an extension to the JavaScript language provided by React. Its convenient syntax is *similar* to HTML and allows you to define complex UI component compositions in a very simple way using its markup.

Now that we have the `TopBar` component ready, let's continue by implementing the application drawer.

Creating the MainDrawer component

The MainDrawer component is used to display navigation links to the different pages and sections of the task manager application. Currently, there are still no pages to link to, so we'll just add a single item pointing to the home page. Whenever we implement a new page, we'll add a new entry to the list. The resulting component should look like this when rendered in its *expanded* state:

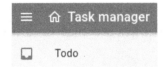

Figure 7.6 – A screenshot of the expanded MainDrawer component

The drawer can also be in a collapsed state. In that case, the items won't show the page name, and a tooltip will be presented when the user focuses or hovers over the icon button. The resulting component should look like this when rendered in its *collapsed* state:

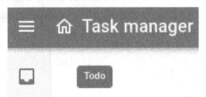

Figure 7.7 – A screenshot of the collapsed MainDrawer component

Let's create a new file in the `src/layout` directory called `MainDrawer.js` that will contain the component's source code. The following snippet contains the relevant part of the code:

```
export const MainDrawer = ({drawerOpen, toggleDrawer}) => (
  <Drawer
    open={drawerOpen} onClose={toggleDrawer}
variant='permanent'
    sx={{
      width: theme => drawerOpen ? theme.layout.drawerWidth :
theme.spacing(7),
        '& .MuiDrawer-paper': theme => ({
          width: theme.layout.drawerWidth,
          ...(!drawerOpen && {
            width: theme.spacing(7),
            overflowX: 'hidden'
          })
```

```
      })
    }}
  >
    <Toolbar/>
    <Box sx={{overflow: drawerOpen ? 'auto' : 'hidden'}}>
      <List>
        <Item disableTooltip={drawerOpen} Icon={InboxIcon}
          title='Todo' to='/'/>
      </List>
    </Box>
  </Drawer>
);
```

Just like we did for the TopBar component, in the first line, we're defining a named export (export const MainDrawer) and assigning the variable with an arrow function that implements a React functional component. In this case, the component accepts two properties – toggleDrawer, which we already analyzed, and drawerOpen. The drawerOpen property accepts a Boolean value to indicate whether the drawer is collapsed or expanded. The component's implementation is also using a composition of several MUI components. Let's study the most relevant ones:

- <Drawer>: In Material Design, a navigation drawer provides access to destinations in your application, so <Drawer> is the MUI component used to define them. There are several variants available – in our case, we'll use the permanent variant. This kind of drawer sits on the same surface elevation as the rest of the content and can be open (visible) or closed (hidden). However, we'll tweak the component's style by setting its sx attribute so that the drawer is never completely hidden, but collapsed.

  In the code snippet, we're setting the sx.width and sx.'& .MuiDrawer-paper' properties of Drawer with a function that accepts the MUI theme. The sx property can either be set with a fixed value or with a function that receives the theme (this can be useful if you want to calculate a value based on a theme property). In our case, if the drawer is open, we return the standard drawer width; if it's not, a width equivalent to seven spacing units is used.

- <Toolbar>: We've already learned about the Toolbar component and its usage within AppBar. However, in this case, we're using it as a spacer. Since the TopBar component sits on top of the MainDrawer component, we use Toolbar to push the rest of the drawer's components down so that they appear just beneath the TopBar component and not behind it.

- <Box>: This component serves as a wrapper and can be used to provide styling. In this case, we're using it to configure the overflow whenever the drawer is collapsed to prevent a scroll bar from being shown.

- <List>: In Material Design, lists are continuous, vertical indexes of text or images. They are composed of items that enable users to perform the actions represented by the element's text description or image. In our case, we'll use this component to wrap the <Item> entries with links to each of the application's pages.

The List component contains an <Item> entry that we haven't analyzed yet. This is another of the components that we need to define within the MainDrawer.js file. The following snippet contains the code for this component:

```
const Item = ({Icon, iconSize, title, to,
disableTooltip=false}) => {
  const match = Boolean(useMatch(to));
  return (
    <ListItemButton component={Link} to={to}
      selected={match}>
      {Icon && <Tooltip title={title} placement='right'
        disableHoverListener={disableTooltip}>
        <ListItemIcon><Icon fontSize={iconSize}/>
          </ListItemIcon>
      </Tooltip>
      }
      <ListItemText primary={title}/>
    </ListItemButton>
  )
};
```

Just like the rest of the components we've defined, this is a React functional component that accepts the following properties:

- Icon: The icon image to display in the list item
- iconSize: The size variant for the icon image
- title: The text to display in the list item when the drawer is expanded, or in the tooltip when collapsed
- to: The target page
- disableTooltip: Whether the tooltip should be disabled when the drawer is collapsed

The component function implementation starts by calling the useMatch React hook provided by React Router to verify that the current browser location matches the target page. We will use this to highlight the navigation <Item> that points to the current user's location.

The function returns a composition of several components. Let's analyze those we haven't checked yet:

- `<ListItemButton>`: This MUI component intends to wrap each of the clickable entries of `<List>`. The `component` attribute allows you to specify what the effective root component of the entry will be. In our case, we're setting it to a React Router `<Link>` component. The `selected` attribute is used to mark the entry as selected; in our case, it will be selected whenever the location matches the item's target. The rest of the attributes will be passed on to the `Link` component.

- `<Link>`: This React Router component renders an HTML hyperlink whose destination targets the effective path denoted by the `to` attribute.

- `<ListItemIcon>`: This MUI component wraps around the icon to be displayed in the list item entry.

- `<ListItemText>`: This MUI component wraps around the text to be displayed in the list item entry. It allows you to set specific properties for the text node. In this case, we're setting the `primary` textual content of the element.

React hooks

React hooks were initially introduced in React 16.8, and allow users to use state and other features without writing a class. Before hooks, users who needed to provide state or life cycle management for their components had to implement a JavaScript class. With hooks, users can define their components using standard JavaScript functions and still take advantage of these features.

Now that we have created the required components for the application's layout, let's learn how to globally manage the layout's state.

Managing the layout's state

In this chapter, we're creating the main layout for our application. The only feature we're adding so far is the possibility to expand and collapse the drawer. This could be easily achieved using a React hook within the layout component. However, in the next few chapters, we'll be adding additional functionality that will require a global application state for the layout.

In the *Adding state management* section, we introduced Redux Toolkit and added the required dependencies. Redux Toolkit manages a global store for the application that can be divided into separate slices for better organization. In this section, we'll create a slice for the layout. Let's create a new file in the `src/layout` directory called `redux.js` that will contain the layout-related state source code. The following snippet contains the relevant part of the code:

```
const layoutSlice = createSlice({
  name: 'layout',
```

```
  initialState: {
    drawerOpen: true,
  },
  reducers: {
    toggleDrawer: state => {
      state.drawerOpen = !state.drawerOpen;
    }
  }
});


export const {
  toggleDrawer
} = layoutSlice.actions;
export const {reducer} = layoutSlice;
```

To create the slice, we use the `createSlice` function provided by Redux Toolkit to initialize the `layoutSlice` variable. This function requires three parameters:

- Name: Defines the name of the slice that will be used to prefix the redux action constant names.

- `initialState`: Accepts an object containing the initial state values. In this case, we're creating an object with the `drawerOpen` property, which will be used to evaluate whether the drawer should be expanded or collapsed. We're setting its initial value to `true`, which means that the drawer will be expanded by default.

- `Reducers`: This object contains each of the reducers applicable for this slice. In Redux, a **reducer** is a function that receives the current state and an action object. The function's implementation contains the logic to be applied to the provided state based on the given action to return a new state.

  In this case, we're defining a single reducer, `toggleDrawer`, that will toggle the value of the `drawerOpen` state property.

Once we've created the slice and stored it in the `layoutSlice` constant, we can export the generated `reducer` and action creator functions. Now that we have all of the necessary parts to complete the `Layout` component, let's continue by analyzing its implementation.

Creating the Layout component

We will be using the `Layout` component as a wrapper in each of the application's page definition components to ensure a consistent look and feel.

Let's create a new file in the src/layout directory called Layout.js that will contain the component's source code. The following snippet contains the relevant part of the code:

```
export const Layout = ({children}) => {
  const navigate = useNavigate();
  const dispatch = useDispatch();
  const drawerOpen = useSelector(state => state.layout.
    drawerOpen);
  const doToggleDrawer = () => dispatch(toggleDrawer());
  return (
    <Box sx={{display: 'flex'}}>
      <TopBar
        goHome={() => navigate('/')}
        newTask={() => {/* TODO */}}
        toggleDrawer={doToggleDrawer} drawerOpen=
          {drawerOpen}
      />
      <MainDrawer
        toggleDrawer={doToggleDrawer} drawerOpen=
          {drawerOpen}
      />
      <Box sx={{flex: 1}}>
        <Toolbar />
        <Box component='main'>
          {children}
        </Box>
      </Box>
    </Box>
  );
};
```

In the first line, we're assigning a functional component to a named export Layout. In this case, the only property is children. In React, this is a special property that contains the child elements enclosed by the opening and closing tags of the component. This fits in the powerful composition model on which React is based; when we want to apply this layout to a given page, we simply need to enclose the components of the page with the <Layout> and </Layout> tags. The composition approach is recommended by the React guidelines instead of component inheritance or other more complex strategies.

In the first part of the component implementation, we are consuming some hooks and defining functions. Let's check them out in detail:

- `useNavigate()`: This is a React Router hook that enables programmatic navigation to a given path within the application. By calling this hook, we initialize the navigate variable, which is a function we can later call within our component's event handling attribute functions.

- `useDispatch()`: This is a Redux hook that returns a reference to the `dispatch` function. This function can be used later to dispatch actions to our application's Redux store. In this case, we're using it to define a function called `doToggleDrawer` that we pass to the `TopBar` and `MainDrawer` components. When invoked, the drawer will collapse and expand alternately for each invocation.

- `useSelector(state => state.layout.drawerOpen)`: This is a Redux hook that allows you to extract data from the application's Redux store. In this case, we're using it to query whether the drawer is currently expanded or collapsed. The resulting value is stored in a variable and then passed on to the child components so that they display accordingly.

The return statement of the function is a component composition containing `TopBar`, `MainDrawer`, and `Box` enclosing the child elements provided by the `children` property. We learned about these components when we analyzed the `TopBar` and `MainDrawer` implementations.

The `Layout` component is now ready to be used. Now, let's learn how to implement a dummy page so that we can put everything together before we implement the real application pages in the following chapters.

Displaying a dummy page

In this section, we'll be implementing a temporary dummy page that will allow us to put everything we've learned until now together and see the application in action. To implement this page, we'll create a new file called `InitialPage.js` in the `src` directory. The following snippet contains the relevant parts of the code:

```
export const InitialPage = () => (
  <Layout>
    <Typography variant='h4' >
      Greetings Professor Falken!
    </Typography>
  </Layout>
);
```

InitialPage is just a functional React component that leverages our Layout component to display a message. When we implement the task manager's pages, we'll be using a similar approach and enclosing all of the required page components between the Layout elements.

The page is now ready; however, we still need to provide some additional global settings such as creating the global store, overriding the MUI theme, and setting up the application router before we can navigate to this page.

Setting up the global store

The main requirement to be able to use Redux in our application to obtain global state management is to create a store. Fortunately, Redux Toolkit provides a very simple way to construct one. Let's start by creating a new file called store.js under the src directory, which is where we'll provide its implementation. The following snippet contains the relevant parts of the code (you can find the complete code in this book's GitHub repository at https://github.com/PacktPublishing/Full-Stack-Quarkus-and-React/blob/main/chapter-07/src/main/frontend/src/store.js):

```
const appReducer = combineReducers({
    layout: layoutReducer
});

const rootReducer = (state, action) => {
    return appReducer(state, action);
};

export const store = configureStore({
    reducer: rootReducer
});
```

In the first block, we're creating the global application reducing function (appReducer) by combining the reducing functions for each of the store slices. The Redux combineReducers helper function accepts an object whose properties are assigned to different reducing functions and returns a single reducing function. This is useful when the application grows in complexity to break up the store into different sections that can be managed independently. In this case, we only have a single reducer (layoutReducer), but whenever we implement a new store slice, we'll add its reducer as a new entry.

In the second block, we define the rootReducer reducer function, which we'll use when creating the application's store to define the root reducer. At the moment, it's redundant to the appReducer variable since we're delegating the call to that function. However, in the future, this function will become handy when we want to respond to actions that affect the *whole* application's state and not only one of its slices.

In the third and final block, we are defining the application's store with a named export. We're using Redux Tookit's `configureStore` function for this purpose. This is a simplification of the Redux `createStore` function, which provides opinionated defaults so that we can create a store without the need to supply repetitive configuration. In this case, we're just configuring the `rootReducer` function we created in the previous steps of the code for the `store.js` file.

Now that we have the application's store ready, let's learn how to customize the task manager by providing an extended MUI theme.

Overriding the MUI theme

MUI comes with a predefined theme that includes a nice set of defaults that might be good enough for most use cases. However, if you want to override one of those default settings, you can easily achieve this by providing your own customized theme.

We'll start by creating a new `styles` directory beneath the `src` folder, which will hold the source code files related to the application styling. Next, we'll create a new `theme.js` file in this directory, where we'll implement our theme. The following snippet contains the relevant part of the code (you can find the complete code in this book's GitHub repository at `https://github.com/PacktPublishing/Full-Stack-Development-with-Quarkus-and-React/tree/main/chapter-07/src/main/frontend/src/styles/theme.js`):

```
export const theme = createTheme({
  layout: {
    drawerWidth: 240,
  }
});
```

We are leveraging the `createTheme` function provided by MUI to create a new theme and share it through a named export. The `createTheme` function creates a new theme based on the default theme overriding its defaults with the values provided by the object passed as the first argument. In this case, we're overriding the default drawer's width to make it wider (240 px).

Now that we have set up the global store and overridden some of the MUI theme defaults, let's learn how to set up the application routing.

Setting up the application router

In the *Creating a React project* section, we added the React Router dependency and learned how a frontend routing library can be useful to create bookmarkable and user-friendly URLs for your application. In this section, we'll learn how to set it up in our project.

We'll start by opening the App.js file in our IDE and replacing its code with the following snippet (only the relevant part is shown here; please check this book's GitHub repository for the complete source code at https://github.com/PacktPublishing/Full-Stack-Development-with-Quarkus-and-React/tree/main/chapter-07/src/main/frontend/src/App.js):

```
export const App = () => (
  <BrowserRouter>
    <Routes>
      <Route exact path='/' element={<Navigate to='
        /initial-page' />} />
      <Route exact path='/initial-page' element={
        <InitialPage />} />
    </Routes>
  </BrowserRouter>
);
```

With these changes, we have modified the purpose of this component to make it the main router for our application. React Router is extremely simple to use since it provides several components that you can combine to create the routes for the application. Let's analyze what each component does:

- `<BrowserRouter>`: This component is the standard implementation of React Router's `Router` interface. React Router provides several router implementations from which you can choose, depending on your needs and your application's nature: `NativeRouter`, `HashRouter`, and so on. `BrowserRouter` is the recommended implementation and the one that'll allow us to provide user-friendly URLs.

- `<Routes>`: This is a required component for enclosing one or more `Route` component definitions.

- `<Route>`: This component defines each of the router entries. Each `Route` contains a matcher expression in its `path` attribute and a React component reference in the `element` attribute. If the current browser location matches one of the route entries path expressions, React Router will render the component defined in the `element` property of the matching route.

  Our first `Route` definition matches the root path and uses the `Navigate` component to redirect the user to the `/initial-page` path. The second `Route` definition matches the `/initial-page` path and renders the `InitialPage` component. Considering both routes, each time a user navigates to the `/` URL, React Router will redirect the user to the `/initial-page` path and eventually render the `InitialPage` component.

Now that we have implemented the initial page, the application's global store, a specific theme, and configured the routing, let's learn how to put everything together and test that everything works as expected.

Putting it all together

The last step before we can launch the application is putting all of the pieces we just created together. We'll need to edit the `src/index.js` file and replace its content with the following code (only the relevant part is shown here; please check this book's GitHub repository for the complete source code at `https://github.com/PacktPublishing/Full-Stack-Development-with-Quarkus-and-React/tree/main/chapter-07/src/main/frontend/src/index.js`):

```
const root = ReactDOM.createRoot(document.
getElementById('root'));
root.render(
  <React.StrictMode>
    <Provider store={store}>
      <ThemeProvider theme={theme}>
        <CssBaseline />
        <App/>
      </ThemeProvider>
    </Provider>
  </React.StrictMode>
);
```

The `ReactDOM.createRoot` and `root.render` statements are common to most React applications and are used to render the application in the `root` HTML DOM element. In the `root.render` method argument, we pass the components we want React to render. Let's learn what each of them does:

- `<React.StrictMode>`: `StrictMode` is a React development tool that highlights potential problems in the application and helps you identify them by logging them into the console. These checks are only active when the application runs in development mode.

- `<Provider store={store}>`: This component is provided by Redux to make the store available to its descendent components. In this case, we're providing access to the store we created in the *Setting up the global store* section.

- `<ThemeProvider theme={theme}>`: This MUI component allows you to customize the theme of its descendants. If you don't have a custom theme, this is *optional*. In our case, we are using it to provide the theme we created in the *Overriding the MUI theme* section.

- `<CssBaseline>`: This MUI component performs a global CSS reset to provide a simple, elegant, and consistent style baseline across browsers. When rendered, it will add the required styles to the web page.

- `<App/>`: This is the actual component that React will render. Since we changed it to define the route definitions for our application, different components will be rendered, depending on the browser location.

In the following screenshot, you can see the resulting code and the directory tree listing all of the files we've created and modified throughout this chapter:

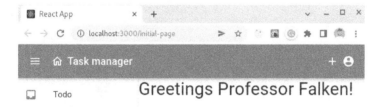

Figure 7.8 – A screenshot of IntelliJ displaying the content of the index.js file

The application's dummy page should be ready by now. Let's try to execute our application and load it in a browser. We'll start by executing the following command:

```
npm start
```

The development server process should start and we should be able to navigate to the application's home. Let's try to check the application in a browser by navigating to `http://localhost:3000`:

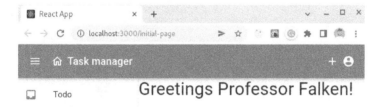

Figure 7.9 – A screenshot of a browser pointing to http://localhost:3000/initial-page

The browser should automatically redirect us to `http://localhost:3000/initial-page` and we should see the result of rendering the `InitialPage` component. Notice how all of the layout components render appropriately and how the sample content we provided on this page is placed in the main area. We can try to click on the menu icon and see how the drawer collapses and expands alternately. By leveraging React's powerful composition model, creating new pages that share this same look and feel will be effortless.

Summary

In this chapter, we learned how to bootstrap the frontend application for our project. We started by learning about React and its advantages. Then, we learned how to create a React project and added the required dependencies to support the task manager features. Next, we created a common layout to provide a consistent look and feel for our application. Finally, we created an initial dummy page to put everything together and be able to see our application in action.

You should now have a good overview of React and the required dependencies to be able to bootstrap a functional frontend application from scratch. In the next chapter, we'll set up the frontend application so that we can consume the secured backend HTTP API. We'll create a login page to initialize the authentication flow, and the required logic to protect some of the URLs and routes.

Questions

1. Who created Material Design and what was its purpose?

2. Why should we use a frontend routing solution?

3. Does React prescribe any routing library?

4. What are the advantages of Redux Toolkit compared to the standard Redux library?

5. What is a React hook?

6. What is JavaScript XML?

7. What do the React guidelines recommend – composition or inheritance?

8
Creating the Login Page

In this chapter, we'll implement the authorization infrastructure for the frontend side of our task manager application, and create login and user management pages to test it. We'll start by configuring both Quarkus and React to be able to consume the secured HTTP APIs in development mode. Then, we'll learn how to use Redux Toolkit to manage the frontend session, and create a login page to be able to initiate a new one. Finally, we'll create a protected user management page, and run the application to verify that everything works as expected.

By the end of this chapter, you should be able to use Redux Toolkit to implement the authentication and authorization infrastructure for React applications to be able to consume secure HTTP APIs based on JWT.

We will be covering the following topics in this chapter:

- Setting up the authentication workflow
- Creating the Login page
- Adding a protected user management page
- Running the application

Technical requirements

You will need the latest Java JDK LTS version (at the time of writing, this is Java 17). In this book, we will be using Fedora Linux, but you can use Windows or macOS as well.

You will need the latest **Node.js** LTS version (at the time of writing, this is 16.15).

You will need a working Docker environment to deploy a PostgreSQL database and to create a Linux native image using Docker. There are Docker packages available for most Linux distributions. If you are on a Windows or macOS machine, you can install Docker Desktop.

You can download the full source code for this chapter from `https://github.com/PacktPublishing/Full-Stack-Quarkus-and-React/tree/main/chapter-08`.

Setting up the authentication workflow

In *Chapter 4*, *Securing the Application*, we learned how to implement a security layer in Quarkus using JWT, and then used it to protect the backend side of the task manager. Attempting to consume the HTTP API from the frontend without an adequate authentication workflow and infrastructure would result in forbidden and unauthorized responses. In this chapter, we'll see how to create an auth service that will enable us to perform login requests to obtain a JWT and use it to authorize the requests to protected endpoints.

Now, let's learn how to set up Quarkus to be able to consume its API from a React application running in dev mode.

Configuring Quarkus for React dev mode

When we deliver our application to a production environment, both the React frontend side and the HTTP API will be served by Quarkus. However, during the development phase, we'll be running the React application in dev mode using the `npm start` command we analyzed in *Chapter 7*, *Bootstrapping the React Project*. The HTTP requests we perform from the frontend application to the Quarkus backend will originate from React's development server at `localhost:3000` and target the Quarkus development server at `localhost:8080`. Your browser's built-in **cross-origin resource sharing** (**CORS**) policy will very likely block these requests, so we need to configure Quarkus' CORS filter to allow this origin.

We can do this by editing the `src/main/resources/application.properties` file and adding the following properties:

```
%dev.quarkus.http.cors=true
%dev.quarkus.http.cors.origins=http://localhost:3000
```

Notice that both properties are prefixed with `%dev.`, which enables them only for Quarkus dev mode. Let's see what each of these configuration properties does:

- `%dev.quarkus.http.cors`: This property enables the Quarkus CORS filter. The filter will identify cross-origin HTTP requests and add the required headers to allow cross-origin access if applicable.

- `%dev.quarkus.http.cors.origins`: This property sets the list of allowed origins that can perform cross-origin requests. In this case, we're setting the URL where we'll be accessing the frontend application.

Now that we've configured Quarkus, let's continue by configuring React.

> **CORS**
>
> CORS is a mechanism that allows a server to share resources to origins (domain, scheme, or port) different from its own, which would normally be prevented by a browser, by providing special HTTP headers.

Configuring React's environments

Depending on the target environment, development or production, the frontend application will be served from different servers. When we run the application in development mode, it will be served from `http://localhost:3000`. When we run the application in production mode, it will be served from the same URL as the backend. This is a problem since the HTTP API will be available at different locations, depending on the environment. To overcome this, we'll be using environment variables defined in `.env` files.

We'll create the following files and content:

- `src/main/frontend/.env`:

  ```
  REACT_APP_API_URL=/api/v1
  ```

- `src/main/frontend/.env.development`:

  ```
  REACT_APP_API_URL=http://localhost:8080/api/v1
  ```

In both files, we're defining `REACT_APP_API_URL`, which we'll consume later on to compute the URL where we'll perform the HTTP requests. For the development environment, the URL includes the domain and port since the backend and frontend will be served separately. Note that the environment variables we define on these files will be embedded at build time into the final static application code bundle (the HTML, CSS, and JS files).

> **Note**
>
> In *Chapter 7, Bootstrapping the React Project*, we initialized the application in the `src/main/frontend` directory. All of the directories and files referenced in the rest of this chapter related to the frontend application modules will be relative to this folder.

With that, we have configured Quarkus and React to be able to consume the backend HTTP API from the frontend application. Now, let's continue by implementing the frontend session management functionality.

Managing the frontend session

We'll use Redux Toolkit and our application's global store to perform the required HTTP requests to obtain the JWT and to manage the user's session.

Redux makes it straightforward to manage the application's global state and to perform mutations through actions and reducers. However, it gets harder when the actions that should mutate the state are *delayed* or *asynchronous*, or if they require information from the application's store. A typical example is when you need to query an HTTP endpoint and wait for the delayed response to get back and then perform an action, depending on the response status or its content. **Redux Thunk**, a component of the Redux ecosystem, is the standard way to write asynchronous logic within Redux-based applications. Redux Toolkit includes the Thunk component and several additional functions that help create thunks with less boilerplate code.

We'll start by creating an `auth` directory beneath the `src` folder. This is where we'll create all of the auth-related services and components. Then, we'll create a new `redux.js` file in this directory where we'll implement the code to manage the auth state. Now, let's analyze some of the code snippets in this file (you can find the complete code in this book's GitHub repository at `https://github.com/PacktPublishing/Full-Stack-Development-with-Quarkus-and-React/tree/main/chapter-08/src/main/frontend/src/auth/redux.js`):

```
export const login = createAsyncThunk(
  'auth/login',
  async ({name, password}, thunkAPI) => {
    const response = await fetch(`${process.env.
      REACT_APP_API_URL}/auth/login`, {
      method: 'POST',
      headers: {
        'Content-Type': 'application/json',
      },
      body: JSON.stringify({name, password}),
    });
    if (!response.ok) {
      return thunkAPI.rejectWithValue({
        status: response.status, statusText: response.
          statusText, data: response.data});
    }
    return response.text();
  }
);
```

In the previous piece of code, we're initializing a `login` Redux action creator function that takes advantage of the `createAsyncThunk` function provided by Redux Toolkit. Notice that Redux Toolkit provides an even simpler approach to fetching data from a remote HTTP API that we'll take advantage of for the rest of the modules. However, the login logic has some additional complexities that require us to use this more involved approach.

In the first parameter, we're providing the `'auth/login'` value, which will be used as a prefix to create the Redux action constant names (since we're using Redux Toolkit, which deals with Redux configuration for us, these constant names are not important and remain transparent to us; we just need to make sure they are unique).

In the second argument, we're providing the `payloadCreator` function, an asynchronous function that performs the login logic. The first parameter of this function is the argument that will be passed to the resulting action creator function – in our case, the user's credentials. The second parameter, `thunkApi`, is an object provided by Redux Toolkit that can be used to access the Redux store functions (`getState` and `dispatch`) or to complete the action by *fulfilling* or *rejecting* it.

In our `payloadCreator` function implementation, we start by performing a POST request to the login endpoint we defined in *Chapter 4*, *Securing the Application*. Notice how we use the REACT_APP_API_URL environment variable to get the base URL of our backend server. If the request fails, we return the result of invoking the `rejectWithValue` function with the response details. If the request is successful, we return the promise that was received when invoking the `response.text()` function.

Now, let's continue by implementing the store slice that will manage the authentication state. The following code snippet contains the most relevant part:

```
const authSlice = createSlice({
  name: 'auth',
  initialState: {
    jwt: sessionStorage.getItem('jwt')
  },
  reducers: {
    logout: state => {
      sessionStorage.removeItem("jwt");
      state.jwt = null;
    }
  },
  extraReducers: {
    [login.fulfilled]: (state, action) => {
      sessionStorage.setItem('jwt', action.payload);
      state.jwt = action.payload;
```

```
        }
      }
  });
```

In the *Managing the layout's state* section of *Chapter 7, Bootstrapping the React Project*, we learned about Redux Toolkit's `createSlice` function and how the global application's store can be divided into separate slices for better organization. In the preceding snippet, we're creating a new slice to store the auth-related data, namely the JWT token that holds the user's session. However, to be able to keep the token even if the user refreshes the browser page, we'll be storing the JWT in the browser's session storage too.

For the `initialState` values, we try to load the token from the browser's storage API: `sessionStorage.getItem('jwt')`. In the `reducers` object, we define a `logout` reducer that clears the JWT both from the application's state and the browser's session storage.

For this slice, we're providing an additional `extraReducers` configuration. This object allows a slice to respond to actions that *don't* belong to itself. In this case, we're configuring a reducer that responds to the result of the action of the `login` thunk we created previously in this section. The reducer logic persists the JWT obtained from a *successful* (fulfilled) request both in the browser's session storage and in the application's state.

We have now fully implemented the Redux part for the auth module. However, we're going to export an additional function in the `src/auth/redux.js` file that will be used by the rest of the application's modules to perform *authorized* HTTP requests. The following snippet contains the code for this function:

```
export const authBaseQuery = ({path}) => {
  const baseQuery = fetchBaseQuery({
    baseUrl: `${process.env.REACT_APP_API_URL}/${path}`,
    prepareHeaders: (headers, {getState}) => {
      headers.set('authorization', `Bearer ${getState().
        auth.jwt}`);
      return headers;
    }
  });
  return async (args, api, extraOptions) => {
    const result = await baseQuery(args, api,
      extraOptions);
    if (result.error && result.error.status === 401) {
      api.dispatch(logout());
    }
```

```
      return result;
   };
};
```

authBaseQuery is a function that wraps the fetchBaseQuery function provided by Redux Toolkit, which, in turn, is a wrapper around the standard JavaScript fetch function, which simplifies the process of preparing and performing HTTP requests.

In the first part, we're invoking fetchBaseQuery with our predefined base URL and including the JWT as a Bearer authorization HTTP request header. Notice how we retrieve the token for the header value using the provided getState() function to access the auth slice and the current jwt. In the *Letting users change their password* section of *Chapter 4, Securing the Application*, we learned how to authorize the requests in cURL using the bearer authentication. The resulting HTTP request dispatched by the authBaseQuery function should be similar to those.

In the second part, we're wrapping the original baseQuery function to be able to inspect the HTTP response before passing it on to the reducer. If the response is not successful, then we dispatch a logout() action. In case we wanted to implement some additional functionality to renew expired tokens in the future, this would be a good place to fit in that logic. If the response is successful, then we just return and pass on the original result. From now on, whenever we need to perform an *authorized* HTTP API request, we can use the authBaseQuery convenience function we just defined instead of the regular fetchBaseQuery function.

The redux part of the auth module is ready. Next, we need to set it up in our task manager application store. Let's edit the src/store.js file to do so. The following code snippet contains the relevant parts of the required changes:

```
const appReducer = combineReducers({
   auth: authReducer,
   layout: layoutReducer
});

const rootReducer = (state, action) => {
   if (logout.match(action)) {
      state = undefined;
   }
   return appReducer(state, action);
};
```

In the first block, we're adding the reducer for the auth module slice to the global appReducer by adding a new entry to the combineReducers function. We'll repeat the same process for each of the slices we implement in the future.

In the second block, we're adding logic to clear the *complete* application's state in case a `logout` action is dispatched. In the *Setting up the global store* section of *Chapter 7, Bootstrapping the React Project*, we initiated `rootReducer` but we didn't implement any functionality. The purpose of this function is now clearer – whenever we need to react to actions that affect the whole application state, this is the place where we'll locate the required code. The alternative would be to implement similar logic in the `extraReducers` section of each slice, but that would be repetitive and hard to maintain.

We've now completed the required functionality to create and manage a user session in React. Let's continue by implementing a login page to be able to start the authentication workflow.

Creating the Login page

The first step of initiating the authentication workflow for our application is to request a JWT from the backend by performing a login request. For this purpose, in our task manager application, we'll create a login page where users will be redirected whenever they don't have an active session or their token has expired. The resulting page should look like this when rendered:

Figure 8.1 – A screenshot of the rendered Login page

To implement the component, let's create a new file in the `src/auth` directory called `Login.js`, which will contain its source code. Let's analyze some of the relevant parts (you can find the complete code in this book's GitHub repository at `https://github.com/PacktPublishing/Full-Stack-Quarkus-and-React/blob/main/chapter-08/src/main/frontend/src/auth/Login.js`):

```
export const Login = () => {
  const dispatch = useDispatch();
  const navigate = useNavigate();
  const {values, isValid, error, setError, onChange} =
    useForm({
    initialValues: {username: '', password: ''}
  });
```

```
const sendLogin = () => {
  if (isValid) {
    dispatch(login({name: values.username,
      password: values.password}))
      .then(({meta, payload}) => {
        if (meta.requestStatus === 'fulfilled') {
          navigate('/');
        } else if (payload?.status === 401) {
          setError('Invalid credentials');
        } else {
          setError('Error');
        }
      });
  }
};
```

We start by defining and exporting a React functional component named Login that contains a component composition to render the form shown in *Figure 8.1*. Next, we initialize the dispatch and navigate variables from their corresponding React Router and React Redux hooks, just like we learned in the *Creating the Layout component* section of *Chapter 7, Bootstrapping the React Project*.

Then, we initialize several variables and functions using a useForm hook. You can check its complete implementation in this book's GitHub repository in the src/useForm.js file. This hook provides common functionality when dealing with forms in our application. It's a combination of React hooks that allow you to manage the state of the form data through the values and onChange variables, and also its validation and error reporting.

In the last part, we're defining a sendLogin function. This function contains the logic that will be processed when the user presses the **SIGN IN** button. In the implementation, first, we check whether the form data is valid. If it's valid, we dispatch the result of the login action creator function invocation with the credentials from the form data. We also subscribe to the asynchronous result of the dispatch invocation. If the result is fulfilled, we redirect the user to the home page (navigate('/')). If it isn't, we set the applicable error message.

Now, let's focus on the return statement of the login page. The following snippet contains its code:

```
return (
  <Container maxWidth='xs'>
    <Box sx={{mt: theme => theme.spacing(8), display:
      'flex', flexDirection: 'column', alignItems:
        'center'}}>
      <Avatar sx={{m: 1}}>
```

```
          <LockOutlinedIcon />
        </Avatar>
        <Typography component='h1' variant='h5'>
          Sign in
        </Typography>
        <Box noValidate sx={{ mt: 1 }}>
          <TextField margin='normal' required fullWidth
            autoFocus
            label='Username' name='username' onChange=
              {onChange} value={values.username}
          />
          <TextField type='password' margin='normal' required
            fullWidth
            label='Password' name='password' onChange=
              {onChange} value={values.password}
            onKeyDown={e => e.key === 'Enter' && sendLogin()}
          />
          <Button fullWidth variant='contained'
            onClick={sendLogin} sx={{ mt: 3, mb: 2 }}>
            Sign In
          </Button>
        </Box>
      </Box>
      <Snackbar
        open={Boolean(error)} message={error}
        autoHideDuration={6000} onClose={() =>
          setError(null)}
      />
    </Container>
  );
```

We've already learned about some of these components; let's study the most relevant and those we haven't covered yet:

- `<Avatar>`: In Material Design, avatars are a type of imagery that focuses on a single subject that represents a user, a brand, or a single entity. `<Avatar>` is the MUI component used to define them. In our case, we're using it as a handy way to display a rounded lock icon.

- `<TextField>`: This MUI component lets users enter and edit text. In our case, we're using two types: a regular text field and a password text field that obscures the user input. Both are *controlled* components that read their data from the `values` variable we obtained from the `useForm` invocation. Notice that the `name` attribute value must match the name of the variable where the data is persisted so that the `onChange` function works accordingly.

- `<Button>`: In Material Design, buttons are used by users to perform actions and make choices with a single click or tap. `<Button>` is the MUI component used to define them. In our case, we're using the button to trigger the `sendLogin` function and start the login process.

- `<Snackbar>`: In Material Design, Snackbars are used to display temporary short messages near the bottom of the screen. The messages should provide information about a process that the application has performed or that will be performed in the future. `<Snackbar>` is the MUI component used to define them. In our case, we're using it to display any of the error messages produced when dispatching the login action. The following screenshot shows the message that's displayed when introducing invalid credentials:

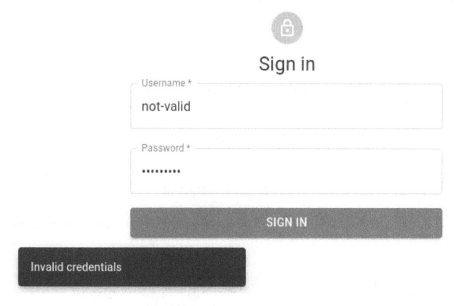

Figure 8.2 – A screenshot of the rendered login page with a visible snackbar

Now that we've defined the login page, we need a way for users to be redirected to this page when they don't have an active session. For this purpose, we'll complete the following steps:

1. Open the `src/App.js` file in your editor and add the following entry to the `<Routes>` children:

```
<Route exact path='/login' element={<Login />} />
```

This defines a new application route called /login, which will display our login page whenever the browser points to this URL.

2.   Open the src/layout/Layout.js file in your editor and add the following lines after the dispatch variable definition:

```
const jwt = useSelector(state => state.auth.jwt);
useEffect(() => {
  if (!jwt) {
    navigate('/login');
  }
}, [navigate, jwt]);
```

Here, we're checking the current application's state to see whether a JWT has been set in the auth slice. If one hasn't, we redirect the user to the /login route, which displays the login page. Since the Layout component is reused for all of the application pages, it's a good way to share this logic and ensure that a user with no session or an invalid one is always redirected to the login page.

So far, we've covered how to manage the user frontend session by persisting the JWT, and how to request it using a login page. However, our application defines two user roles (admin and user), and we still need to make sure that users with a specific role can only access those parts of the application they are authorized to. Now, let's create a user management page to demonstrate how to achieve this.

Adding a protected user management page

In this section, we'll define a user management page for users with the admin role that displays the application's registered users. Let's start by defining the user module API service.

Defining a user API Redux slice using Redux Toolkit's createApi function

We'll begin by creating a users directory beneath the src folder, where we'll create all of the user management-related services and components. Next, we'll create a new api.js file where we'll implement the required functionality. The following snippet contains the most relevant part of the code (you can find the complete code in this book's GitHub repository at https://github.com/PacktPublishing/Full-Stack-Development-with-Quarkus-and-React/tree/main/chapter-08/src/main/frontend/src/users/api.js):

```
export const api = createApi({
  reducerPath: 'users',
  baseQuery: authBaseQuery({path: 'users'}),
  tagTypes: ['User'],
  endpoints: builder => ({
```

```
    getUser: builder.query({
      query: id => `/${id}`,
      providesTags: ['User']
    }),
    getUsers: builder.query({
      query: () => '/',
      providesTags: ['User']
    }),
    deleteUser: builder.mutation({
      query: user => ({
        url: `/${user.id}`,
        method: 'DELETE'
      }),
      invalidatesTags: ['User']
    }),
    getSelf: builder.query({
      query: () => '/self',
      providesTags: ['User']
    }),
    changePassword: builder.mutation({
      query: passwordChange => ({
        url: `/self/password`,
        method: 'PUT',
        body: passwordChange
      })
    })
  })
});
```

createApi is the main function provided by *Redux Toolkit Query* and allows you to define and configure a set of endpoints to retrieve data from an HTTP API and then use it to populate and mutate a Redux store slice. In this case, we're using it to define a slice for users and retrieve data from the backend user resource. Let's analyze the parameters of the configuration object we used in the previous snippet:

- reducerPath: Redux Toolkit uses this value internally to define the unique path within the global application's state where the created slice reducer will be processed. Besides the requirement to provide a unique path, the value is not important and remains transparent to us unless we access the application's state directly.

- baseQuery: This is the base query or HTTP request configuration that will be used to fetch each endpoint. Since these endpoints are secured, we'll be reusing the authBaseQuery function we created in the *Managing the frontend session* section, which adds the required authorization headers for us. In the path configuration argument, we're providing a users value, which means that the resulting base URL will be ${process.env.REACT_APP_API_URL}/users.

- tagTypes: Redux Toolkit Query provides a caching mechanism for the data retrieved from each endpoint. Tag types are an optional setting that will enable us to persist and invalidate the cache associated with a given type. In our case, we're just setting a generic User tag.

- endpoints: Here, we define each of the server endpoints we want to consume and the kind of operation we want to perform. For those endpoints where we are just retrieving data (read), we use query operations. These only require setting the URL path through the query field, and the tag that the cached data should be associated with through the providesTags field.

  For those operations that perform changes (create, update, or delete), we can use mutation operations. The configuration is very similar to the one for query operations. Here, in addition to the path, we're also providing the HTTP method and the request body if applicable. Notice how the delete endpoint has an invalidatesTags setting, which will remove all the cached data and force Redux Toolkit to re-fetch it.

So far, we've learned that the createApi function defines endpoint-triggering functions and a Redux store slice that holds the state of the result of querying these endpoints. As a final step, we need to configure our application's store to account for this slice by editing the src/store.js file and making the following changes:

```
const appReducer = combineReducers({
  auth: authReducer,
  layout: layoutReducer,
  [userApi.reducerPath]: userApi.reducer
});
// …
export const store = configureStore({
  reducer: rootReducer,
  middleware: getDefaultMiddleware => getDefault
    Middleware()
```

```
      .concat(userApi.middleware)
});
```

In the first part, we're adding the exported slice reducer to the global `appReducer`. In the second block, we're configuring the store's middleware to take the API's middleware into account by adding it to the default set provided by Redux Toolkit. Middleware is an advanced concept that's useful for asynchronous API calls. Luckily enough, Redux Toolkit Query hides the complexity for us and only requires this simple setup.

Now that the user API module is ready, let's create a page to learn how to consume it.

Creating the Users page

The Users page is a very simple page that lists the registered users and their roles in a table and provides an action item to delete them.

To implement this component, we must create a new file in the `src/users` directory called `Users.js`, which will contain its source code. You can find the complete code for this component in this book's GitHub repository at `https://github.com/PacktPublishing/Full-Stack-Quarkus-and-React/blob/main/chapter-08/src/main/frontend/src/users/Users.js`. The implementation is very similar to the other components we've already created. Let's analyze the API consumption part:

```
const {data: allUsers} = api.endpoints.getUsers.useQuery
   (undefined, {pollingInterval: 10000});
const {data: self} = api.endpoints.getSelf.useQuery();
const [deleteUser] = api.endpoints.deleteUser.
   useMutation();
```

The first two statements take advantage of the `useQuery` hook, which triggers the data fetch from an endpoint to initialize some variables. The returned `data` variable will contain the data from the cache if it exists, and the updated data from the endpoint once the response has been processed. In addition, for the first statement, we're configuring a `pollingInterval`, which will force a new fetch every 10 seconds. This means that if new users were created in the backend, the user list would automatically update with the new additions. The last statement takes advantage of the `useMutation` hook. This hook provides a function to dispatch the `deleteUser` endpoint action.

Now, let's check how these variables and functions are used in the following snippet to form the Users page component:

```
{allUsers && allUsers.map(user =>
  <TableRow key={user.id}>
    <TableCell>{user.name}</TableCell>
    <TableCell>{new Date(user.created).toLocaleDateString
      ()}</TableCell>
    <TableCell>{user.roles.join(', ')}</TableCell>
    <TableCell align='right'>
      <IconButton
        disabled={user.id === self?.id} onClick={() =>
          deleteUser(user)}
      >
        <DeleteIcon/>
      </IconButton>
    </TableCell>
  </TableRow>
)}
```

We start by iterating the array of users (allUsers) that was returned by the query hook. For each user, we present the user's name, its creation date, and a comma-separated list of its roles in different table cells. In the last table cell, we provide IconButton that, when clicked, causes the deleteUser function to supply the user as an argument. In addition, this icon is only enabled if the user is not the same as the user performing the query (self). In the following screenshot, you can see what the table looks like when rendered:

Users

Name	Created	Roles	
admin	6/5/2022	admin, user	🗑
user	6/5/2022	user	🗑

Figure 8.3 – A screenshot of the rendered Users table

The page is ready; however, it is still unreachable for users. Just like we did for the login page, we need to add a new entry to the `<Routes>` definition in the `src/App.js` file:

```
<Route exact path='/users' element={<Users />} />
```

This defines a new application route called `/users`, which will display our `Users` page whenever the browser points to this URL.

Now, let's modify `MainDrawer` so that it displays a link for users with the `admin` role.

Adding a link to the Users page in MainDrawer

We want our application to provide a great *user experience*. For that purpose, we want to show a link to the user management page but only if the user has the `admin` role. Let's create a new `HasRole.js` file in the `src/auth` directory to implement a component that will help us achieve this. The following snippet contains the most relevant parts of its content (you can find the complete code in this book's GitHub repository at `https://github.com/PacktPublishing/Full-Stack-Quarkus-and-React/blob/main/chapter-08/src/main/frontend/src/auth/HasRole.js`):

```
export const HasRole = ({role, children}) => {
  const {data} = api.endpoints.getSelf.useQuery();
  if ((data?.roles ?? []).includes(role)) {
    return children;
  }
  return null;
}
```

This component can be used to wrap other components and show them only if the currently logged-in user has the specified `role`. In its implementation, we're taking advantage of the `getSelf` endpoint to get the information of the current user and check whether any of its `roles` match the provided `role`. The child components will be rendered only if there is a match.

Now, we can edit the `src/layout/MainDrawer.js` file so that it includes the new link to the user management page. The following code snippet contains the required changes:

```
<List>
  <Item disableTooltip={drawerOpen} Icon={InboxIcon}
    title='Todo' to='/'/>
  <HasRole role='admin'>
    <Divider/>
    <Item disableTooltip={drawerOpen} Icon={PersonIcon}
```

```
        title='Users' to='/users'/>
    </HasRole>
  </List>
```

The main change is we're adding a new Item entry to link the /users route with the particularity that it's wrapped with a HasRole component with the role='admin' attribute. If everything works as expected, this item will only be visible to admin users.

Now, let's start the application to see all of the changes in action.

Running the application

We're still in the development phase, so we need to start both the frontend and backend applications in dev mode. To start the Quarkus backend, just as we've done previously, we'll execute the following command from the project root:

```
./mvnw quarkus:dev
```

The backend server should start and be ready to serve requests. Next, in a *different* terminal, and from the frontend project location (src/main/frontend), we'll execute the following command:

```
npm start
```

The frontend application should start and a browser window should open automatically at http://localhost:3000. The page should load and automatically redirect us to http://localhost:3000/login:

Figure 8.4 – A screenshot of a browser pointing to http://localhost:3000/login

Now, let's log in with the administrator credentials by entering `admin` in the username field, `quarkus` in the password field, and pressing the **SIGN IN** button. The process should complete successfully and we should be redirected to `http://localhost:3000/initial-page`:

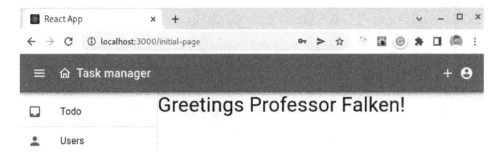

Figure 8.5 – A screenshot of a browser pointing to http://localhost:3000/initial-page for the admin user

Notice how the drawer displays a link to the **Users** page. Let's click on it:

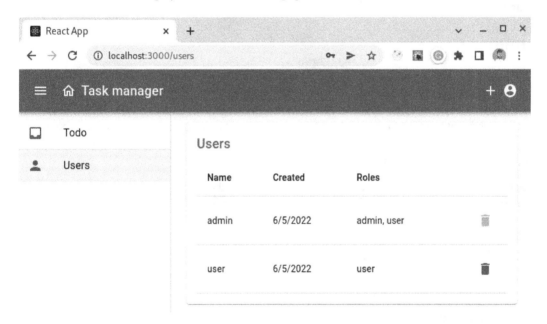

Figure 8.6 – A screenshot of a browser pointing to http://localhost:3000/users

The **Users** page should load, and we should be able to see a table containing the list of registered users. We've now verified that the application works fine for an `admin` user and that you can access and interact with the parts of the task manager that are only intended for this kind of user.

Since we haven't implemented a logout mechanism yet, let's close this browser tab and open a new one and navigate to the application's root: `http://localhost:3000`. We should be redirected to the login page – let's log in, but with a regular user instead. We'll repeat the previous process, but now, we'll use `user` as the username. Once again, the process should complete successfully and we should get redirected to `http://localhost:3000/initial-page`:

Figure 8.7 – A screenshot of a browser pointing to http://localhost:3000/initial-page for the regular user

The layout is the same as for the `admin` user, but now, the link to the **Users** page is missing from the drawer.

Summary

In this chapter, we learned how to configure Quarkus and React to be able to consume the secured HTTP APIs from our application's backend. Then, we used Redux Toolkit to implement the infrastructure to manage the frontend session. Next, we implemented a login page to be able to create new sessions, and a user management page to showcase how to protect components from users who don't have authorized roles. Finally, we learned how to start the backend and the frontend applications in development mode to be able to check that everything works accordingly.

You should now be able to use Redux Toolkit to implement the authentication and authorization infrastructure for React applications to be able to consume secure HTTP APIs based on JWT.

In the next chapter, we'll learn how to implement the main logic of the frontend application. We'll replace the main screen and implement some CRUD operations that will make use of the backend HTTP API.

Questions

1. What is CORS?
2. What's the recommended and easiest approach to consuming an HTTP API and persisting its results in a Redux store?
3. What kind of operations does Redux Toolkit Query support for endpoints?
4. What's a snackbar in Material Design?
5. What's the default URL for React's development server?

9

Creating the Main Application

In this chapter, we'll create the core features for the frontend application, which will complete the task manager we've been developing throughout this book. We'll start by adding user-specific features that will allow users to easily log out from the application and change their passwords. We'll also add the required functionality for users to be able to create projects so that they can group tasks into different categories, and the task management features so that they can create, delete, update, and mark tasks as complete.

By the end of this chapter, you should be able to use Redux Toolkit and MUI to implement user interfaces in React to be able to consume REST HTTP APIs. Being able to create user interfaces for your backend will allow you to create complete usable full-stack applications. You should be able to understand the complete cycle of web application development, from the persistence layer through the backend, and expose it via an HTTP API to the frontend and the final interface so that users can interact with your application.

We will be covering the following topics in this chapter:

- Adding user-specific features
- Adding CRUD functionalities
- Deleting the no longer needed files and running the application

Technical requirements

You will need the latest Java JDK LTS version (at the time of writing, this is Java 17). In this book, we will be using Fedora Linux, but you can use Windows or macOS as well.

You will need the latest **Node.js** LTS version (at the time of writing, this is 16.15).

You will need a working Docker environment to deploy a PostgreSQL database and to create a Linux native image using Docker. There are Docker packages available for most Linux distributions. If you are on a Windows or macOS machine, you can install Docker Desktop.

You can download the full source code for this chapter from `https://github.com/PacktPublishing/Full-Stack-Quarkus-and-React/tree/main/chapter-09`.

Adding user-specific features

In *Chapter 8, Creating the Login Page*, we implemented the Redux store slices and developed the required logic to manage users and their authentication and authorization. In this section, we'll take advantage of these features to implement a dialog for users to be able to change their passwords. We will also add an icon to the `TopBar` component to allow users to perform actions related to their account and session management, such as triggering the change password workflow or logging out of the application. Let's start by implementing the password change dialog.

Implementing a password change dialog

To implement the password change dialog, we'll create a new React component called `ChangePasswordDialog` that will contain two required text fields to introduce – the current and new passwords – and two action buttons to confirm or cancel the action. The resulting component should look like this when rendered:

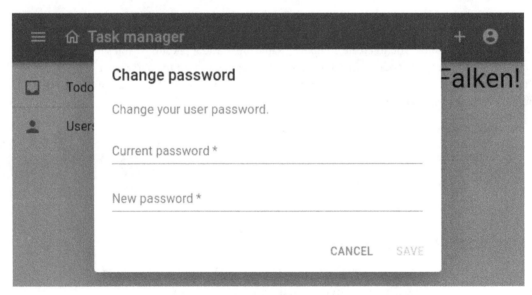

Figure 9.1 – A screenshot of the ChangePasswordDialog component

> **Note**
>
> In *Chapter 7, Bootstrapping the React Project*, we initialized the application in the `src/main/frontend` directory. All of the directories and files referenced in the rest of this chapter related to the frontend application modules will be relative to this folder unless stated otherwise.

To implement the component, let's create a new file in the `src/users` directory called `ChangePasswordDialog.js`, which will contain its code. You can find the complete source code for this file in this book's GitHub repository at `https://github.com/PacktPublishing/Full-Stack-Development-with-Quarkus-and-React/tree/main/chapter-09/src/main/frontend/src/users/ChangePasswordDialog.js`. Let's analyze the most relevant parts here:

```
export const ChangePasswordDialog = () => {
  const {values, invalid, isValid, error, setError,
    clearForm, onChange} = useForm({
    initialValues: {currentPassword: '', newPassword: ''}
  });
  const dispatch = useDispatch();
  const changePasswordOpen = useSelector(state =>
    state.layout.changePasswordOpen);
  const close = () => dispatch(closeChangePassword());
  const [changePassword] = api.endpoints.changePassword.
    useMutation();
  const canSave = isValid &&
    Boolean(values.currentPassword) &&
      Boolean(values.newPassword);
  const save = () => {
    changePassword(values).then(({error}) => {
      if (!Boolean(error)) {
        clearForm();
        close();
      } else if (error?.status === 409) {
          setError('Current password is incorrect');
      } else {
        setError('Unknown error, please try again');
      }
    });
  };
```

We start by defining and exporting a React functional component called `ChangePasswordDialog` that contains a component composition to render the Material Design-based dialog shown in *Figure 9.1*. Since the component contains a form to capture user input, we're reusing the `useForm` hook we described in the *Creating the login page* section of *Chapter 8, Creating the Login Page*, which provides form state management and validation features. The `const canSave` variable declaration uses some of the values returned by our hook to determine whether the form is valid and filled in, and can be saved.

Then, we initialize the `changePasswordOpen` variable, which will be used to determine whether the dialog is visible or not. This variable is read from the global application state and will allow us to open and close the dialog from different parts of the application. This variable should be used in conjunction with the `closeChangePassword` function, which can be dispatched to the application's store to close the change password dialog.

In the *Adding a protected user management page* section in *Chapter 8, Creating the Login Page*, we initialized the user API Redux slice, which contained the `changePassword` endpoint definition. In the last part of the preceding snippet, we declare the `save` function that will send the data to the backend by leveraging the endpoint's mutation hook. The `save` function starts by dispatching the `changePassword` endpoint action with the values retrieved from the form. Once the response from the server is processed, in the `then` function, we check whether it successfully closed the dialog. If it didn't, we use the `setError` function provided by the `useForm` hook to present a relevant error message to the user.

The following screenshot shows how the error message would be rendered in case a user introduced an invalid current password:

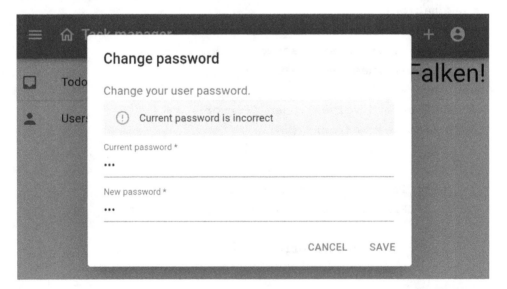

Figure 9.2 – A screenshot of the ChangePasswordDialog component displaying an error message

Now, let's focus on the return statement of the `ChangePasswordDialog` component. The following snippet contains its code:

```
<Dialog open={changePasswordOpen} onClose={close}>
  <DialogTitle>Change password</DialogTitle>
  <DialogContent>
    <DialogContentText>
      Change your user password.
    </DialogContentText>
    {Boolean(error) && <Alert severity='error'>
      {error}</Alert>}
    <TextField type='password' fullWidth margin='dense'
      variant='standard' label='Current password'
      name='currentPassword' value={values.currentPassword}
        onChange={onChange}
      required error={Boolean(invalid.currentPassword)}
      autoFocus
    />
    <TextField type='password' fullWidth margin='dense'
      variant='standard' label='New password'
      name='newPassword' value={values.newPassword}
        onChange={onChange}
      required error={Boolean(invalid.newPassword)}
    />
  </DialogContent>
  <DialogActions>
    <Button onClick={close}>Cancel</Button>
    <Button onClick={save} disabled={!canSave}> Save
    </Button>
  </DialogActions>
</Dialog>
```

We learned about some of these components in *Chapter 7*, *Bootstrapping the React Project*, and *Chapter 8*, *Creating the Login Page*. Let's study the most relevant and those we haven't covered yet:

- `<Dialog>`: In Material Design, dialogs are a type of modal window used to inform users about a task. They usually contain critical information, involve multiple tasks, or require decisions. In our case, we're using the dialog to request the user a new password and perform the password change operation. `<Dialog>` is the MUI component used to define the main dialog container and controls if it's hidden or displayed. The `open` attribute expects a Boolean property indicating if the dialog is displayed or hidden. The `onClose` attribute expects a function that will be invoked when the user clicks or taps on the scrim area (in Material Design, the scrim is the shaded area that appears behind the dialog) and that should complete by closing the dialog.

- `<DialogTitle>`: This MUI component is used to specify the dialog's title that appears in the upper part of the modal dialog window. In our case, we'll just specify a static literal `Change password`.

- `<DialogContent>`: This MUI component encloses the dialog's content that appears in the middle section of the dialog. This is where we'll add the dialog child components required to render the change password form.

- `<Alert>`: This is an MUI-specific component that displays short messages in a way that attracts the user's attention. This is the component that we'll use to render the error messages, as shown in *Figure 9.2*. The component accepts a `severity` configuration attribute to control if the alert is displayed as an informative message, a warning, an error, and so on. In this case, we check if there's an error message to conditionally display this `Alert` with an `error` type severity.

- `<DialogActions>`: This MUI component encloses the action buttons that appear in the bottom part of the dialog. In our case, we're adding a cancel button that's always enabled, and a save button that will only be enabled when the `canSave` condition is true.

Now that we've implemented the dialog component, we need to add it to our layout so that we can display it whenever the user invokes a change password action. Since our layout component is common to every application page, users should be able to change their password from any of the application's pages, provided that they are logged in.

Let's open the `src/layout/Layout.js` file and add the new component entry to the last part of the component definition. You can find the complete source code for this file in this book's GitHub repository. The following snippet shows the affected part only:

```
<Box sx={{flex: 1}}>
  <Toolbar />
  <Box component='main'>
    {children}
  </Box>
</Box>
<ChangePasswordDialog />
```

We added the component right after the main area container of the layout. Despite the component being *always* in the layout composition, it will only be displayed if the application's global state property, `layout.changePasswordOpen`, is true.

With that, we have implemented the dialog and added it to the layout, but we still need a way to display it. Now, let's add an icon to the `TopBar` component so that users can perform actions to manage their accounts.

Adding a user icon to the top bar

So far, we've created the `ChangePasswordDialog` component, which will only be displayed if the `layout.changePasswordOpen` global state property is true. However, we haven't added this property to the layout Redux slice yet. Let's edit the `src/layout/redux.js` file we created in the *Managing the layout's state* section of *Chapter 7, Bootstrapping the React Project*, and make the following changes:

```javascript
const layoutSlice = createSlice({
  name: 'layout',
  initialState: {
    changePasswordOpen: false,
    drawerOpen: true,
  },
  reducers: {
    openChangePassword: state => {
      state.changePasswordOpen = true;
    },
    closeChangePassword: state => {
      state.changePasswordOpen = false;
    },
    toggleDrawer: state => {
      state.drawerOpen = !state.drawerOpen;
    }
  }
});


export const {
  openChangePassword, closeChangePassword, toggleDrawer
} = layoutSlice.actions;
```

Let's analyze these changes in detail:

- `initialState`: This creates a new field called `changePasswordOpen` with a `false` default value. This setting forces the dialog to be hidden by default whenever the application is accessed for the first time.

- `reducers`: This creates two new reducer functions that will be used to change the `changePasswordOpen` property, one to open the dialog (`openChangePassword`) and one to close it (`closeChangePassword`). Alternatively, we could have created a single function with a parameter to set the value.

- `exports`: This includes the newly generated action creator functions in the list of exported constants.

Now that we've completed the layout's Redux slice changes, let's continue by implementing the user icon as a separate component. This icon is a clickable button that displays a pop-up menu with user-related tasks when clicked. The resulting icon and menu should look like this when rendered:

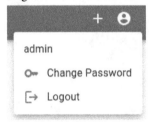

Figure 9.3 – A screenshot of the UserIcon component with its pop-up menu

To implement the component, let's create a new file in the `src/layout` directory called `UserIcon.js`, which will contain its code. You can find the complete source code for this file in this book's GitHub repository at `https://github.com/PacktPublishing/Full-Stack-Quarkus-and-React/blob/main/chapter-09/src/main/frontend/src/layout/UserIcon.js`. Let's analyze the most relevant parts:

```
export const UserIcon = () => {
  const [anchorEl, setAnchorEl] = useState(null);
  const menuOpen = Boolean(anchorEl);
  const closeMenu = () => setAnchorEl(null);
  const dispatch = useDispatch();
  const {data} = api.endpoints.getSelf.useQuery();
```

We start by defining and exporting a React functional component named `UserIcon`, which contains a component composition that renders a clickable icon of a person. This expands a pop-up menu when clicked, as shown in *Figure 9.3*. Next, we initialize the `anchorEl` and `setAnchorEl` variables using React's built-in `useState` hook. This hook is used to preserve a local state within this component;

its invocation returns a variable with the state and a function to be able to mutate it. In this case, we're using it to store the reference to the component that holds the user's context menu. anchorEl holds the reference to this component, which is initially null. The menu should appear closed or hidden whenever this variable is null, which we explicitly define in the menuOpen variable. We also define a convenience function, closeMenu, that encapsulates a call to the setAnchorEl function with a null value that will close the pop-up menu when invoked.

In the last statement, we take advantage of the useQuery hook of the Redux user slice's getSelf endpoint to retrieve the information of the currently logged-in user from the backend. In this case, the main purpose is to use this information to show the currently logged-in user's name as the first entry of the pop-up menu. This same information could be used to create a profile page for the user as a future evolution of the application.

Now, let's focus on the return statement of UserIcon, where we define the component composition. The following snippet contains its code:

```
<Tooltip title='Profile'>
  <IconButton color='inherit' onClick={event =>
    setAnchorEl(event.currentTarget)}>
    <AccountCircleIcon />
  </IconButton>
</Tooltip>
<Menu
  anchorEl={anchorEl}
  open={menuOpen}
  onClose={closeMenu}
>
  {data && <MenuItem>{data.name}</MenuItem>}
  <MenuItem onClick={() => {
    dispatch(openChangePassword());
    closeMenu();
  }}>
    <ListItemIcon>
      <KeyIcon />
    </ListItemIcon>
    Change Password
  </MenuItem>
  <MenuItem onClick={() => dispatch(logout())}>
    <ListItemIcon>
```

```
        <LogoutIcon />
      </ListItemIcon>
      Logout
    </MenuItem>
  </Menu>
```

We've already learned about some of these components. Let's study the most relevant and those we haven't covered yet:

- `<IconButton>`: We learned about this MUI component in the *Creating the TopBar* section of *Chapter 7, Bootstrapping the React Project*. However, let's focus on the `onClick` attribute, which contains the function that will display the pop-up menu. When the user clicks on the button, the function takes the HTML event to set the `anchorEl` value with the **Document Object Model** (**DOM**) element that was clicked.

- `<Menu>`: In Material Design, menus display a list of choices on temporary surfaces; `<Menu>` is the MUI component used to define them. The `anchorEl` attribute contains a reference to an HTML element that will be used by MUI as a reference to position the menu. The `open` attribute specifies whether the menu is visible or not. The `onClose` attribute accepts a function that closes the menu. This function is invoked whenever the user clicks outside the menu or presses the *Esc* key.

- `<MenuItem>`: This MUI component must be used to encapsulate each of the menu entries. It can contain an `onClick` attribute that invokes a function whenever the user clicks on it. In our case, we have three entries. The first entry displays the logged-in user's name. The second opens the **Change password** dialog when clicked. The third dispatches the logout action from the auth Redux slice we created in the *Managing the frontend session* section of *Chapter 8, Creating the Login Page*.

Now that we have implemented and analyzed the `UserIcon` component that will allow the user to open the **Change password** dialog, let's add it to the `TopBar` component. Let's open the `src/layout/TopBar.js` file and add the new component entry to the last part of the component definition. You can find the complete source code for this file in this book's GitHub repository at `https://github.com/PacktPublishing/Full-Stack-Quarkus-and-React/blob/main/chapter-09/src/main/frontend/src/layout/TopBar.js`. The following snippet only shows the affected part:

```
      <UserIcon />
    </Toolbar>
  </AppBar>
```

The changes involve just adding the component as the last element in the `Toolbar` component. This means that when the toolbar is rendered, the new icon will be displayed as the rightmost element.

With that, we've implemented a change password dialog and added a user context menu to the application's layout top bar. We've also made the required changes to the application's Redux store to be able to open and close the new dialog. Now, let's continue by creating some of the task manager application's pages to implement its task and project management functionalities.

Adding CRUD functionalities

The task manager application allows users to create tasks, manage their completion status, assign them different levels of priority, group them into projects, and so on. So far, we've learned how to bootstrap a generic frontend application with global state management and we've added some features related to user management. However, we haven't implemented anything related to the main application functionalities yet. Let's continue by implementing the project CRUD features.

Adding the project management features

The main purpose of projects is to allow the application's users to group the tasks they create. For example, users can create a *Home* and a *Work* project and add home-specific tasks or work-specific tasks to each of them. Later, they can filter the tasks for each of these projects, so they can focus on each of them separately. Our first goal is to provide the means for users to be able to create projects. Let's begin by implementing the required functionality to interact with the HTTP API.

Defining the project API Redux slice

In the *Adding a protected user management page* section of *Chapter 8*, *Creating the Login Page*, we learned about Redux Toolkit's `createApi` function and how we can use it to interact with an HTTP API backend to populate and mutate a Redux store slice. For each of the backend entities (Users, Projects, Tasks, and so on), we will create a Redux Toolkit API implementation.

Let's create a `projects` directory under the `src` directory where we'll create all of the project-related services and components. Next, we'll create a new `api.js` file where we'll implement the required functionality. The following snippet contains the most relevant part of the code. You can find the complete source code for this file in this book's GitHub repository at `https://github.com/PacktPublishing/Full-Stack-Quarkus-and-React/blob/main/chapter-09/src/main/frontend/src/projects/api.js`:

```
export const api = createApi({
  reducerPath: 'projects',
  baseQuery: authBaseQuery({path: 'projects'}),
  tagTypes: ['Project'],
```

```
endpoints: builder => ({
  getProjects: builder.query({
    query: () => '/',
    providesTags: ['Project'],
  }),
  addProject: builder.mutation({
    query: project => ({
      url: '/',
      method: 'POST',
      body: project
    }),
    invalidatesTags: ['Project'],
  }),
  updateProject: builder.mutation({
    query: project => ({
      url: `/${project.id}`,
      method: 'PUT',
      body: project
    }),
    invalidatesTags: ['Project'],
  })
})
});
```

We covered the main `createApi` function options in *Chapter 8, Creating the Login Page*. In this case, we're adding three endpoint definitions to deal with the project HTTP API resource:

- `getProjects`: This endpoint definition queries the `projects` HTTP endpoint, which returns the list of projects for the currently logged-in user.

- `addProject`: This is the endpoint definition for creating a new project. When invoked, it performs an HTTP POST request to the `projects` endpoint with the data provided for the new project, which for now is just the new project's name.

- `updateProject`: This is the endpoint definition for updating an existing project. When invoked, it performs an HTTP PUT request to the `projects` endpoint for the provided project `id` with the updated project data.

For each of the new store slices we define in the application, we'll need to edit the `src/store.js` file to add the slice reducer and middleware to the application's global store.

In the following snippet, we are adding the project API reducer:

```
const appReducer = combineReducers({
  auth: authReducer,
  layout: layoutReducer,
  [projectApi.reducerPath]: projectApi.reducer,
  [userApi.reducerPath]: userApi.reducer
});
```

In the following snippet, we are adding its middleware:

```
export const store = configureStore({
  reducer: rootReducer,
  middleware: getDefaultMiddleware =>
    getDefaultMiddleware()
    .concat(projectApi.middleware)
    .concat(userApi.middleware)
});
```

With that, we've finished implementing the Redux Toolkit API slice for projects. This allows us to interact with the backend through its HTTP API very easily.

Let's continue by implementing some components to deal with projects that will take advantage of this API.

Creating components to deal with projects

In our task management application, projects are used to group and classify the user's tasks. Currently, the Project entity just contains a field for the project's name. In the future, the application might evolve in such a way that projects could offer more functionality and store additional information such as start and end dates, collaborators, and so on. However, given their current simplicity, there's no need to create specific application pages to manage projects.

To create new projects, we'll add a new button in the application drawer that will show a modal dialog where the user can specify the project name and confirm its creation. To assign a task to an existing project, we'll create a drop-down list from which users can select an existing project.

Implementing a new project dialog

To implement the new project dialog, we'll create a new React component called NewProjectDialog that will contain a single text field to introduce the new project name, and two action buttons to confirm or cancel the project creation action. The resulting component should look like this when rendered:

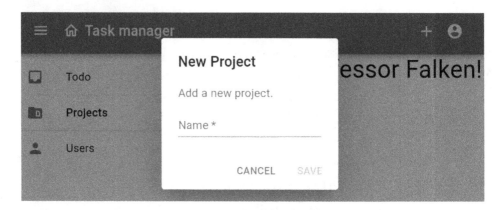

Figure 9.4 – A screenshot of the NewProjectDialog component

The implementation of this component follows the same patterns we learned for `ChangePasswordDialog` in the *Implementing a password change dialog* section. We won't go over them again here; you can find the complete source code for this component at https://github.com/PacktPublishing/Full-Stack-Quarkus-and-React/blob/main/chapter-09/src/main/frontend/src/projects/NewProjectDialog.js.

Following that approach, we'll also need to modify the layout's Redux store slice in the `src/layout/redux.js` file so that it includes a new field called `newProjectOpen` that will be `true` if the modal dialog is visible:

```
initialState: {
  changePasswordOpen: false,
  drawerOpen: true,
  newProjectOpen: false,
},
```

We'll also add two reducer functions to show and hide the modal dialog:

```
openNewProject: state => {
  state.newProjectOpen = true;
},
closeNewProject: state => {
  state.newProjectOpen = false;
},
```

The NewProjectDialog component implementation, which will enable users to create new projects, is ready. In the *Finalizing the application's layout* section, we'll improve the application's drawer to list the projects and allow users to filter their tasks by clicking on them. We'll also add a button so that users can dispatch the openNewProject reducer function to show the dialog and create new projects. Let's continue by implementing another generic component to list and select the user's projects.

Implementing a project selector drop-down list

The main goal of projects is to allow users to group their tasks and filter them. Let's create a drop-down list selector component that will allow users to select a project when they are editing or creating a new task. This component should be generic so that it can be reused on different parts of the application too. When collapsed, the component is just an icon button of a label. When the user presses the button, a pop-up menu is expanded, showing the list of projects for that user. The resulting component should look like this when rendered:

Figure 9.5 – A screenshot of the SelectProject component

To implement this component, let's create a new file in the src/projects directory called SelectProject.js, which will contain its code. You can find the complete source code for this file in this book's GitHub repository at https://github.com/PacktPublishing/Full-Stack-Development-with-Quarkus-and-React/tree/main/chapter-09/src/main/frontend/src/projects/SelectProject.js. Let's analyze the most relevant parts:

```
export const SelectProject = ({disabled, onSelectProject = ()
=> {}}) => {
```

This snippet shows the SelectProject React functional component definition, which has two attributes: disabled and onSelectProject. The disabled attribute can be passed in to enable or disable the icon button that toggles the pop-up menu. The onSelectProject attribute expects a function that will be triggered whenever the user clicks or taps on one of the menu entries. When the onSelectProject function is invoked, it receives an object representing the project as its single argument; the project corresponds with the one the user clicked on from the dropdown of available entries.

In the following snippet, we take advantage of the `useQuery` hook of the Redux project slice's `getProjects` endpoint to retrieve the list of projects for the user:

```
const {data: projects} = api.endpoints.getProjects.
useQuery(undefined);
```

This list is used to populate the pop-up menu entries that are displayed, as shown in *Figure 9.5*. The rest of the component follows a similar pattern as the `UserIcon` component we analyzed in the *Adding a user icon to the top bar* section.

With that, we've implemented the project-related components and Redux store logic. Now, let's continue by implementing the task management-related features and putting everything together.

Adding the task management features

Tasks are the core feature of our task management application. In this section, we'll implement all of the components related to rendering tasks and the required business logic to interact with the backend through its HTTP API. Let's start by implementing the Redux API slice.

Defining the task API Redux slice

Just like we did for the projects in the *Defining the project API Redux slice* section, we'll start by implementing the task API Redux slice that will be used both to manage the state for tasks and to interact with the HTTP API. Let's create a `tasks` directory under the `src` directory, where we'll create all of the task-related services and components. Next, we'll create a new `api.js` file, where we'll implement the required functionality. This implementation is almost identical to what we did for users in the *Defining a user API Redux slice using the Redux Toolkit createApi function* section of *Chapter 8*, *Creating the Login Page*, as well as for the projects in the *Defining the project API Redux slice* section. We won't go over this again here; you can find the complete source code at `https://github.com/PacktPublishing/Full-Stack-Quarkus-and-React/blob/main/chapter-09/src/main/frontend/src/projects/NewProjectDialog.js`.

To complete the slice configuration, we just need to add the reducer and its middleware to the main application's store; then, we'll be done. Let's edit the `src/store.js` file and add the required entries. The following snippet contains the relevant changes:

```
const appReducer = combineReducers({
  /* [...] */
  [taskApi.reducerPath]: taskApi.reducer,
/* [...] */
  middleware: getDefaultMiddleware =>
    getDefaultMiddleware()
    .concat(projectApi.middleware)
```

```
    .concat(taskApi.middleware)
 /* [...] */
```

In this code block, we're adding `taskApi.reducer` to the global `appReducer`, and concatenating `taskApi.middleware` to the global store's `middleware` configuration.

With that, we've finished implementing the Redux Toolkit API slice for tasks. Let's continue by implementing a dialog to display and edit tasks.

Implementing the task edit dialog

To implement the task edit dialog, we'll create a new React functional component named `EditTask` that will contain several fields to edit and display the task data and buttons to save, cancel, and delete the open task. Unlike the other dialogs we've implemented so far, this one will be a full-screen dialog instead of a modal one. The resulting component should look like this when rendered:

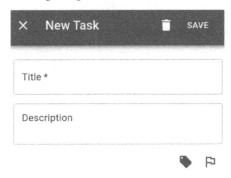

Figure 9.6 – A screenshot of the EditTask component when creating a new task

To implement the component, let's create a new file in the `src/tasks` directory called `EditTask.js`, which will contain its code. You can find the complete source code for this file in this book's GitHub repository at `https://github.com/PacktPublishing/Full-Stack-Quarkus-and-React/blob/main/chapter-09/src/main/frontend/src/tasks/EditTask.js`. Let's analyze the most relevant parts:

```
 const openTask = useSelector(state => state.layout.openTask);
```

In the previous code block, we're retrieving the currently open task from the global application state. For this to work, we need to modify the layout's Redux slice to add an `openTask` field. Since this follows the same approach we already covered for the change password dialog in the *Implementing a password change dialog* section, we won't go over this again. You can find the code for the final layout redux implementation at `https://github.com/PacktPublishing/Full-Stack-Quarkus-and-React/blob/main/chapter-09/src/main/frontend/src/layout/redux.js`.

Continuing with the `EditTask` component implementation, in the following snippet, we are defining two variables based on the currently open task:

```
const isNew = openTask && !Boolean(openTask.id);
const isComplete = openTask && Boolean(openTask.complete);
```

The `isNew` variable is used to select the kind of operation to perform when saving – `addTask` in case it's true or `updateTask` in case it's false. It is also used to determine the dialog's title – that is, **New Task** or **Edit Task**. In the following screenshot, we can see the `EditTask` dialog component when editing an existing task:

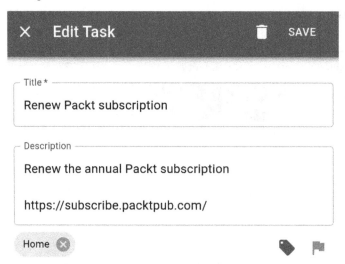

Figure 9.7 – A screenshot of the EditTask component when editing an existing task

The `isComplete` variable will be true in case the task was marked as completed by the user. We will use this to disable the input fields and save button in case the task was completed since we want to prevent users from editing these tasks. In the following screenshot, you can see the same task displayed in *Figure 9.7* once the user marks it as complete:

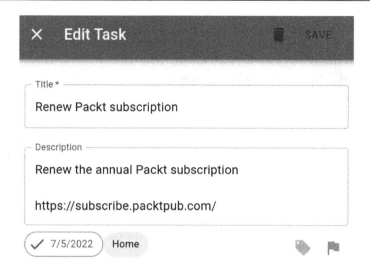

Figure 9.8 – A screenshot of the EditTask component when displaying a complete task

Now, let's take a closer look at the save function, which will be invoked whenever the user presses the save button. The following code snippet contains its implementation:

```
const [addTask] = api.endpoints.addTask.useMutation();
const [updateTask] = api.endpoints.updateTask.useMutation();
const save = event => {
  event.preventDefault();
  if (event.currentTarget.checkValidity()) {
    const operation = isNew ? addTask: updateTask;
    operation(openTask).then(({error}) => {
      if (!Boolean(error)) {
        close();
      }
}
    });
  }
};
```

We start by defining two constants with the possible operations the save function can perform by leveraging the useMutation hooks from the task API Redux slice endpoints we defined in the *Defining the task API Redux slice* section. We'll use the addTask endpoint in case the edited task is new, and the updateTask endpoint in case we're editing an already existing task. In the save function body, we start by validating the form and then invoking the save operation function promise. If the promise succeeds, we close the dialog.

The dialog also contains an icon button of a *trash can*. When the user presses this button, the doDeleteTask function will be invoked. The following snippet contains its implementation:

```
const [deleteTask] = api.endpoints.deleteTask.useMutation();
const doDeleteTask = () => {
  deleteTask(openTask).then(({error}) => {
    if (!Boolean(error)) {
      close();
    }
  })
};
```

This code is very similar to the code we implemented for the save function. However, in this case, we're not performing any kind of validation. We just invoke the deleteTask function, which returns a Promise, and we close the dialog in case it succeeds.

Now, let's analyze the component composition on the return statement of the EditTask component. Since the code is quite long, please check it at https://github.com/PacktPublishing/Full-Stack-Quarkus-and-React/blob/main/chapter-09/src/main/frontend/src/tasks/EditTask.js.

We've already learned about some of these components. Let's study the most relevant and those we haven't covered yet:

- <Dialog>: We covered MUI's Dialog component in the *Implementing a password change dialog* section. However, notice how in this case, we're providing an additional fullScreen attribute, which will configure the dialog to fully cover the visible area.

- <AppBar> and <Toolbar>: In the *Creating the TopBar* section of *Chapter 7, Bootstrapping the React Project*, we introduced both the AppBar and Toolbar components provided by MUI. When we first introduced modal dialogs, we saw that the action buttons should be positioned in the lower area. However, when dealing with full-screen dialogs, Google's Material Design guidelines state that the action buttons should be positioned in the top bar. In our case, we're adding a close icon button, the dialog title, and the delete and save buttons, which will be positioned to the right.

- `<Grid>`: This MUI component allows users to position elements following the Material Design responsive layout grid guidelines. In this case, we use it to position the form elements so that they have consistent spacing and alignment.

- `<CompleteChip>`: This is a custom component that encapsulates an MUI Chip. In Material Design, chips are compact elements that represent an input, attribute, or action. `<Chip>` is the MUI component to define them. You can find the complete implementation at `https://github.com/PacktPublishing/Full-Stack-Quarkus-and-React/blob/main/chapter-09/src/main/frontend/src/tasks/CompleteChip.js`. The `<CompleteChip>` component renders the completion date of the provided task or nothing if the task hasn't been completed yet. The resulting component should look like this when rendered with a completed task:

Figure 9.9 – A screenshot of the CompleteChip component

- `<ProjectChip>`: This is a custom component that encapsulates an MUI Chip to render the provided task's project name. You can find the complete implementation at `https://github.com/PacktPublishing/Full-Stack-Quarkus-and-React/blob/main/chapter-09/src/main/frontend/src/tasks/ProjectChip.js`. The resulting component should look like this when rendered with a task that has been assigned to a project:

Figure 9.10 – A screenshot of the ProjectChip component

- `<SelectProject>`: This is the custom component we implemented in the *Implementing a project selector drop-down list* section. This generic component is used to allow users to select the project to which they want to assign the currently edited task.

- `<EditPriority>`: This is a custom component that allows users to specify the priority of the task. The implementation of this component is very similar to the one we did for the `SelectProject` component; it includes an icon button and a popup menu to select from the list of priority entries. You can find the component's source code at `https://github.com/PacktPublishing/Full-Stack-Quarkus-and-React/blob/main/chapter-09/src/main/frontend/src/tasks/Priority.js`. The resulting component should look like this when rendered:

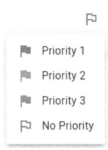

Figure 9.11 – A screenshot of the EditPriority component

We've seen a good overview of the task edit dialog. This is the most complex component of the application; it's used in the application to edit and display information about the user's tasks. You should now have a good understanding of a wide variety of MUI components and be able to create your own by leveraging the composition pattern. Now, let's create a new page using our common Layout component to display a list of tasks.

Implementing a task list page

In the *Displaying a dummy page* section of *Chapter 7, Bootstrapping the React Project*, we implemented a very simple page so that we could start the application and log into it. This page had no real purpose besides having a landing page when the user authenticates and is no longer of any use. Let's replace it with a multipurpose task lister page, which we'll use to display the user's tasks filtered by several criteria.

To implement the task list page, we'll create a new React functional component named Tasks that will reuse our common Layout component to display a list of tasks filtered by different criteria. Depending on how the page is accessed and configured, the filter will display different tasks based on the resulting criteria. The resulting page should look like this when rendered to display the pending, or todo, tasks:

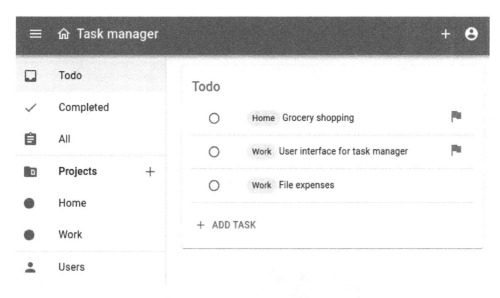

Figure 9.12 – A screenshot of the tasks pending page

To implement the page component, let's create a new file in the `src/tasks` directory called `Tasks.js`, which will contain its source code. You can find the complete code for the page component in this book's GitHub repository at `https://github.com/PacktPublishing/Full-Stack-Quarkus-and-React/blob/main/chapter-09/src/main/frontend/src/tasks/Tasks.js`. Now, let's analyze the most relevant parts:

```
export const Tasks = ({title = 'Tasks', filter = () => true})
=> {
```

The previous snippet contains the functional component declaration. The `title` attribute is used to override the title, which is displayed on top of the table. If nothing is provided, `Tasks` will be used as default. In *Figure 9.12*, the rendered component has a `title` attribute configured with a `Todo` value. The `filter` attribute accepts a function that can be used to filter the tasks. For example, the `t => Boolean(t.complete)` function would configure the page to display only the completed tasks.

In addition to these attributes, the page component can also be configured through React Router URL params:

```
const {projectId} = useParams();
const {project} = projectApi.endpoints.getProjects.
useQuery(undefined, {
   selectFromResult: ({data}) => ({project: data?.find(p => p.id
=== parseInt(projectId))})
});
```

```
if (Boolean(project)) {
  title = project?.name;
  filter = task => task.project?.id === project.id;
}
```

In the first statement, we use React Router's useParams hook to retrieve projectId, which might have been passed through the page URL. In the next statement, we use the useQuery hook provided by the Redux project API slice's getProjects endpoint to retrieve the project that matches projectId from the application's Redux store. The useQuery hook accepts a selectFromResult query option that can be configured with a function. In our case, we compare the IDs of the retrieved projects with the ones we obtained from the URL parameters. If a project matching this ID is found, then we override the title and filter attributes to configure them with the project-specific data. In the following screenshot, you can see what the page looks like when the tasks are filtered by the **Home** project:

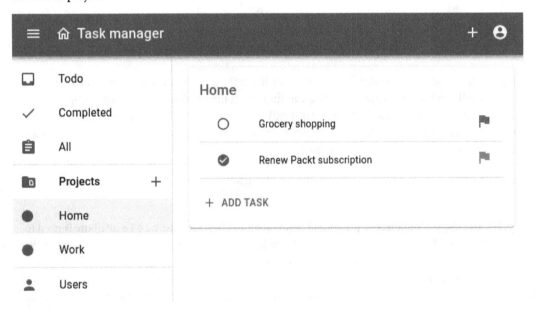

Figure 9.13 – A screenshot of the Home project tasks page

To retrieve the data to populate the table, we consume the useQuery hook from the getTasks endpoint of the task API Redux slice:

```
const {data} = api.endpoints.getTasks.useQuery(undefined,
{pollingInterval: 10000});
```

We configure the hook to poll the data from the backend HTTP API every 10 seconds. The user will always have the most recent data available since the page will update transparently in case changes are received from the backend that update the main application's store.

The return statement is a component composition that displays the table with the filtered tasks. We covered all of the components and how to use them in the *Creating the Users page* section of *Chapter 8, Creating the Login Page*, so we won't go over them again.

The page component is ready. However, we still need to configure the main application router so that the page is rendered whenever the user accesses the applicable URLs.

Adding the task pages to the application router

In the *Setting up the application router* section of *Chapter 7, Bootstrapping the React Project*, we developed the first iteration of our application's router with routes for the login page and the dummy initial page. Now, let's edit the `src/App.js` file to add the definitive routes for the task manager. The following snippet contains the final route configuration with the relevant changes:

```
<Routes>
  <Route exact path='/' element={<Navigate to='/tasks/pending'
/>} />
  <Route exact path='/login' element={<Login />} />
  <Route exact path='/tasks' element={<Tasks />} />
  <Route exact path='/tasks/project/:projectId' element={<Tasks
/>} />
  <Route exact path='/tasks/pending'
        element={<Tasks title='Todo' filter={t => !Boolean(t.
complete)} />} />
  <Route exact path='/tasks/completed'
        element={<Tasks title='Completed' filter={t =>
Boolean(t.complete)} />} />
  <Route exact path='/users' element={<Users />} />
</Routes>
```

Let's see how each route behaves, depending on its configured path attribute:

- `/`: This is the path for the landing page where the users are redirected once they log into the application. Since we want users to focus on their pending tasks, we'll configure this route to redirect them to the pending tasks page. In the element attribute, we configure a React Router `Navigate` component that will take care of the redirection to the applicable path.

- `/login`: This is the route for rendering the `Login` page. Users won't navigate to this page manually, but they'll be redirected if they haven't logged in or their session has expired. The element attribute is configured with the `Login` page component we implemented in the *Creating the Users page* section of *Chapter 8, Creating the Login Page*.

- `/tasks`: This is the route that renders an unfiltered list of tasks. Notice how the element attribute contains an unconfigured `Tasks` page component that will render the complete, unfiltered list of tasks for the current user.

- `/tasks/project/:projectId`: This is the route that renders a list of tasks filtered by a project, as shown in *Figure 9.13*. In this case, the element attribute is also configured with a plain `Tasks` component with no attributes. However, the URL contains a `:projectId` param definition that allows users to enter URLs such as `http://localhost:3000/tasks/project/10`, where `10` would be the project ID passed on to the `Tasks` page. The `Tasks` page component will read this parameter and infer the `filter` and `title` configurations based on the found project.

- `/tasks/pending`: This is the route that renders a list of the user's pending tasks. The element attribute contains a `Tasks` component configured with a `title` and `filter` that shows only the tasks that aren't complete.

- `/tasks/completed`: This is the route that renders a list of the user's completed tasks. The element attribute contains a `Tasks` component configured with a `title` and `filter` that shows only the tasks that have been marked as complete.

- `/users`: This is the route that renders a page containing the list of registered users in the application.

Now that the application router configuration is ready, users should be able to navigate freely to any of the routes we've defined. However, to improve the experience, let's add some links to the application drawer.

Finalizing the application's layout

The `MainDrawer` component's purpose is to display navigation links to the different pages and sections of the task manager application. Let's edit the `src/layout/MainDrawer.js` file to add some links to the routes we just created. You can find the complete set of changes in this book's GitHub repository. Let's analyze the most important ones:

```
<Item disableTooltip={drawerOpen} Icon={InboxIcon} title='Todo'
to='/tasks/pending'/>
<Item disableTooltip={drawerOpen} Icon={CheckIcon}
title='Completed' to='/tasks/completed'/>
<Item disableTooltip={drawerOpen} Icon={AssignmentIcon}
title='All' to='/tasks'/>
```

```
<Projects
  drawerOpen={drawerOpen} openNewProject={openNewProject}
projects={projects}
/>
```

The previous snippet contains the new entries for the drawer. The first three are new Item definitions with links to the pages listing pending, complete, and all tasks. Notice how the to attributes are configured with the routes we defined in the App.js main router for each of these pages.

The last entry in the snippet references a custom Projects component defined in the same MainDrawer.js file. This component accepts a list of projects and creates an Item entry for each project with its corresponding path, as shown in the following snippet:

```
{Array.from(projects).map(p => (
  <Item
    key={p.id} disableTooltip={drawerOpen}
    Icon={CircleIcon} iconSize='small'
    title={p.name} to={`/tasks/project/${p.id}`}/>
))}
```

In this case, the to attribute is configured with a path matching the /tasks/project/:projectId route for each project ID.

In the *Implementing a new project dialog* and *Implementing a task edit dialog* sections, we implemented a modal dialog to create new projects and a full-screen dialog to edit and create tasks. To be able to use them, we need to add them to the application's Layout component. Let's open the src/layout/Layout.js file and add the new component entries to the last part of the component definition. The following snippet shows the required changes:

```
</Box>
<EditTask />
<NewProjectDialog />
<ChangePasswordDialog />
```

We'll also need to add the required attributes to the TopBar and MainDrawer component entries:

```
<TopBar
  goHome={() => navigate('/')}
  newTask={() => dispatch(newTask())}
  toggleDrawer={doToggleDrawer} drawerOpen={drawerOpen}
/>
<MainDrawer
```

```
        toggleDrawer={doToggleDrawer} drawerOpen={drawerOpen}
        openNewProject={doOpenNewProject} projects={projects}
    />
```

The application's layout is now complete. We've added navigation links to all of the application's routes in the main drawer, and configured the common layout to be able to display the dialogs to create projects and edit tasks. Now, let's clean up some of the residual files and start the application to see all of the changes in action.

Deleting the no longer needed files and running the application

We've finished implementing the task manager frontend application functionality. However, we bootstrapped the application using the Create React App script, which added some residual files we no longer need and that we should delete. Let's go ahead and delete the following files from the project:

- src/App.css
- src/logo.svg
- src/InitialPage.js

Now, let's start the application. In the *Running the application* section of *Chapter 8, Creating the Login Page*, we already went through these steps:

1. Start the Quarkus backend from the project root by executing the following command:

    `./mvnw quarkus:dev`

2. In a different Terminal, and from the frontend root (src/main/frontend), start the React development server by executing the following command:

    `npm start`

The frontend application should start and a browser window should open automatically at http://localhost:3000. The page should load and automatically redirect us to http://localhost:3000/login.

After logging in, the application should redirect us to the **Todo** tasks page at http://localhost:3000/tasks/pending:

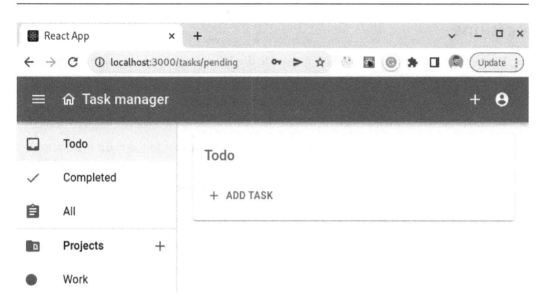

Figure 9.14 – A screenshot of a browser pointing to http://localhost:3000/tasks/pending for a regular user

Let's try to create a new task by pressing the **ADD TASK** button. The `EditTask` dialog should open and we should see an empty form, similar to what's shown in *Figure 9.6*. Let's add some details for this task and press the **SAVE** button. The dialog should close and we should be able to see the new task listed. Now, let's press on the circle to the left of the task to mark it complete. The task should disappear from the **Todo** list but should be visible if we navigate to the **Completed** tasks page.

Everything should be working by now, and a fully functional task management application should be ready. You can play around with the application to check all of the features we implemented: creating new tasks, changing your user password, logging in and out, creating new projects, managing users, and so on.

Summary

In this chapter, we worked on the core features of the frontend application and finished implementing the task manager we've been creating throughout this book. We started by adding user-specific features to allow users to log out of the application and change their passwords. Then, we worked on the core features of the application by adding functionality for users to be able to create projects and perform task management tasks. We also cleaned up the project of residual, not needed files, and learned how to run it in development mode to check that the new functionality is working.

You should now be able to use Redux Toolkit, React Router, and MUI to implement your own React components to build user interfaces that consume REST APIs. In the next chapter, we'll introduce Jest and other frontend testing tools, and learn how to implement tests to ensure that the implemented functionality works according to our specifications.

Questions

1. What's the most appropriate MUI component to display short messages in a way that attracts the user's attention?

2. What's the main purpose of React's built-in `useState` hook?

3. Where are the action buttons located in a modal dialog according to the Material Design guidelines?

4. What additional steps are required to add a Redux Toolkit slice to the main application's Redux store?

5. How do you configure a React Router route that accepts parameters?

10

Testing Your Frontend

In this chapter, we'll implement tests to verify the task manager's frontend features we implemented in the previous chapters. We'll start with an overview of what you need to test JavaScript frontend applications. Then, we'll learn how to implement tests for your React components and verify that our frontend routing solution and its configuration work as expected. Next, we'll implement tests that focus on verifying the application's features as experienced by its users and learn how to run the test suite from the command line.

By the end of this chapter, you should be able to implement unit and integration tests for your JavaScript frontend applications based on React, React Router, and Redux Toolkit. Providing tests for your applications will help you build solid and reliable software and minimize its maintenance effort in the future.

We will be covering the following topics in this chapter:

- Testing frontend applications overview
- Testing React components
- Testing the router
- Testing the application's features
- Running the tests from the command line

Technical requirements

You will need the latest **Node.js** LTS version (16.15 at the time of writing).

You can download the full source code for this chapter at `https://github.com/PacktPublishing/Full-Stack-Quarkus-and-React/tree/main/chapter-10`.

Testing frontend applications overview

The main requirement when testing frontend applications based on JavaScript is having an engine that runs the tests for you. There are plenty of test runner alternatives to choose from – Mocha, Ava, and Jazmine, just to name a few. However, in *Chapter 7, Bootstrapping the React Project*, we created our frontend project by leveraging *Create React App*, which uses *Jest* as its test runner, and this is the one we'll be using for our project. Jest is one of the most popular and widely adopted JavaScript testing frameworks. It's not only a test runner, but a full-featured framework with features and tooling for mocking, assertion, verification, and so on.

One of the crucial parts of frontend testing is verifying that the page elements render correctly in the browser. This is especially critical with React since this library is all about components and managing their state. **Enzyme** was one of the main tools to test the React components' output and rendering until a few years ago. However, this tool is no longer actively maintained and lacks built-in support for the latest React versions, so it should be avoided. Create React App automatically includes a dependency to **React Testing Library** in our project; this is the de facto tool nowadays and the one we'll be using.

React Testing Library was initially released in 2018 as an alternative to Enzyme. It's built on top of the DOM Testing Library, which allows you to query the resulting **Document Object Model** (**DOM**) when rendering your components to interact with them and verify their properties. Verifying individual DOM object behavior as compared to verifying complete snapshots of the page makes your tests more maintainable, robust, and less coupled with the implementation details of your components. Enzyme-based tests are usually focused on React component implementation details, such as their properties or internal state, which makes them brittle and often fail when the tested components are refactored. The main advantage of React Testing Library's approach is that your tests won't break each time you do minor changes to your component or its implementation while still verifying its correct behavior as expected by the end users.

As you can see, most of the required dependencies to test your components were already included by Create React App when we bootstrapped our application. Nonetheless, there are some parts of the code base that would benefit from additional dependencies to reduce the boilerplate when implementing their tests.

In the *Adding state management* section in *Chapter 7, Bootstrapping the React Project*, we introduced *Redux Toolkit* in our project. We could implement tests for the Redux store with the included tools, however, these tests would require cumbersome mocking and would be hard to maintain. Let us now add some test dependencies to our project that will help us with these tasks and allow us to write simpler and more maintainable tests for our application.

Adding the required dependencies

The usage of Redux Toolkit has allowed us to implement a global application state management system that interacts with the backend's HTTP API with very little effort. The easiest way to implement tests that verify the behavior of our application and components that rely on Redux Toolkit Query is by leveraging an HTTP API mocking library. In our case, we'll be using **Mock Service Worker (MSW)**.

Mock Service Worker

MSW is a JavaScript testing library that is used to intercept HTTP requests on the network level and define mock HTTP responses. Since the interception is performed at the network level, it's very easy to integrate into any application testing setup and requires no extra configuration.

To add the MSW dependency, we'll navigate to our frontend application directory (`src/main/frontend`) and execute the following command:

```
npm install --save-dev --save-exact msw@0.43.1
```

The command should complete successfully, and the following lines should be visible in the dependencies section of our `package.json` file:

```
"devDependencies": {
    "msw": "0.43.1"
}
```

Since this dependency is only required for testing purposes, we'll install it as a development dependency in our project. This is achieved by passing the `--save-dev` flag. Notice the usage of the `--save-exact` flag, too. This forces npm to make the `package.json` reference use an exact version rather than a Semver range expression. The resulting reference by default would be `"^0.43.1"` if no argument was passed. Note that the version includes a caret; this would usually imply that whenever the user runs `npm install` on this project, either the `0.43.1` version or any new minor version would be installed. Nonetheless, since this dependency has a `0` major version, npm has a special behavior and installs the latest patch version instead. However, by using an exact version, we make sure that `0.43.1` is *always* installed, regardless of the newer versions available.

Note

In *Chapter 7, Bootstrapping the React Project*, we initialized the application in the `src/main/frontend` directory. All of the directories and files referenced in the rest of this chapter related to the frontend application modules will be relative to this folder unless stated otherwise.

We've learned that frontend JavaScript testing requires a test runner and that verifying that the page elements render and behave as expected is one of the main purposes of these tests. We've gone over the required dependencies and configured our project with the missing ones. Let us now continue by implementing the tests for our application. There are several kinds of tests we can implement depending on their scope, and we'll start by testing individual React components.

Testing React components

Depending on the complexity of your components and their purpose, it might be advisable that you implement specific unit tests for them. This ensures the component behaves according to specification for every defined scenario and property.

> **Unit testing**
>
> Unit testing is a software development technique by which developers write automatic tests that verify that the smallest testable units of an application (called units) behave according to its design requirements.

Let us see a practical example by implementing a test for the `CompleteChip` component. In the *Implementing a Task Edit dialog* section in *Chapter 9, Creating the Main Application*, we created this custom component that renders the completion date of the provided task, or nothing if the provided task is not completed. To implement the test for this component, we'll start by creating a new `CompleteChip.test.js` file in the `src/tasks` directory.

> **Test file naming**
>
> By default, Jest, with the configuration provided by Create React App, searches for tests in files with the `.js` extension under any `__tests__` directory, or files with a `.test.js` extension. For our tests, we'll be using the latter approach, so that we can keep the test files close to the production code they verify.

You can find the complete source code for this file at `https://github.com/PacktPublishing/Full-Stack-Quarkus-and-React/blob/main/chapter-10/src/main/frontend/src/tasks/CompleteChip.js`. Let us analyze the following most relevant parts:

```
describe('CompleteChip component test', () => {
    //...
});
```

The test implementation starts by declaring a Jest `describe` block. These blocks are used to group test cases that are related to each other. It's not required, but it helps keep things organized and also allows the inclusion of setup and teardown code blocks that are common for these groups of tests. In this case, we're declaring the `CompleteChip` component test suite, which includes tests to verify how the component renders when there's no provided task, the provided task is incomplete, and the provided test is completed. The first argument for the `describe` function is the name of the test group, and the second contains a function where you should declare the nested tests and preparation code blocks.

The following code snippet contains the test to verify that `Chip` won't be rendered in case the component is configured with an `undefined task` attribute:

```
test('no task, not visible', () => {
  // When
  const {container} = render(<CompleteChip />);
  // Then
  expect(container).toBeEmptyDOMElement();
});
```

The code for the test starts with a Jest `test` block (you can use its alias, `it`, alternatively). Each test case should be implemented within these blocks where the first argument is the name of the test, and the second contains a function with the test execution code and expectations.

> **Given-When-Then (GWT) / Arrange-Act-Assert (AAA)**
>
> We tend to organize our tests using a GWT or an AAA structure. The first part, Given/Arrange, sets the test context by preparing the original state or environment for the tested scenario. The When/Act phase contains the action that is being tested, and it should contain a single statement in most cases. The final part, Then/Assert, contains the test expectations and assertions that verify the tested action behaves accordingly.

The *When* section of the test renders a `<CompleteChip />` component with no `task` attribute. For this purpose, we use React Testing Library's `render` function that renders the component into an HTML `div` container, which is appended to the DOM's document body. The function returns a container variable with a reference to the `div` DOM node that contains the rendered HTML component.

The last part of the test contains a Jest `toBeEmptyDOMElement` expectation to verify that the `container` is empty just like the component specification requires. This expectation is part of React Testing Library's **jest-dom** Jest extension package. You can find the complete list of expectations at `https://github.com/testing-library/jest-dom`.

The following snippet contains the code to verify that `Chip` is rendered when `CompleteChip` is configured with a completed task:

```
test('complete task, shows date', () => {
  // When
  const {container} = render(<CompleteChip task={{complete:
    '2015-10-21T04:29:00.000Z'}}/>);
  // Then
  expect(container).not.toBeEmptyDOMElement();
```

```
expect(screen.getByText(/2015/)).toBeInTheDocument();
expect(screen.getByTestId('complete-chip')).
  toHaveClass('MuiChip-colorSuccess');
});
```

The `test` action renders a `CompleteChip` component with a task with a valid completion date initializing the HTML DOM `container` variable. The assertion section verifies that the container is not empty. This is the opposite of what was asserted in the previous test case; notice how this is achieved just by prepending `.not` to the `.toBeEmptyDOMElement()` expression.

The next part uses React Testing Library's `screen` functions to query HTML nodes from the entire `document.body` to verify that the React component is rendered accordingly. Let us see these functions in more detail as follows:

- `screen.getByText`

  This function tries to find a component containing the provided text literal or regular expression, throwing an exception if none is found. In our case, we're testing the document to find a component with a text containing `2015`. We're using a regular expression, which implies that the component might contain other text and characters too.

- `screen.getByTestId`

  A function that tries to find an HTML node with a `data-testid` attribute with the provided ID as the value, failing with an exception if none is found. Test IDs are a convenient way to identify unique components for tests, however, they require changes in production code to work. In our case, we need to edit the `src/tasks/CompleteChip.js` file and make the following changes:

  ```
  <Chip
    icon={<CheckIcon />}
    color='success'
    label={new Date(task.complete).toLocaleDateString()}
      variant='outlined'
    data-testid='complete-chip'
  />
  ```

  In the `expectation` section, we're verifying that the `Chip` component is rendered with the success color variant by checking that it has the `MuiChip-colorSuccess` class, using jest-dom's `toHaveClass` matcher.

We have now implemented enough tests to verify that the `CompleteChip` component behaves according to its specifications. Let us now see how to execute these tests in IntelliJ.

Running the tests on IntelliJ

To run the `CompleteChip.test.js` test suite in IntelliJ, we need to click on the play button near the describe block and click on the **Run 'CompleteChip component test'** menu entry.

```
CompleteChip.test.js ×

 4                                                           n: () => {
 5      ▶ Run 'CompleteChip component test'   ▶  Ctrl+Shift+F10
 6      ✿ Debug 'CompleteChip component test'
 7      ⌖ Run 'CompleteChip component test' with Coverage          );
 8        Modify Run Configuration...
 9           expect(container).toBeEmptyDOMElement();
10      });
11     test( name: 'incomplete task, not visible', fn: () => {
12        // When
13        const {container} = render(<CompleteChip task={{title: 'title'}} />);
14        // Then
15        expect(container).toBeEmptyDOMElement();
16     });
```

Figure 10.1 – A screenshot of IntelliJ's Run 'CompleteChip component test' menu entry

The tests should execute and pass, and we should be able to see the results in the **Run** tool window.

```
Run:   CompleteChip component test (1) ×                                                          ✿ —
▶  ✔ ⊘ ↓≡ ↓≡ ⤲ ÷ ↑ ↓ ⊕ ⊾ ⊿ ✿  ✔ Tests passed: 3 of 3 tests – 55 ms
      ✔ Test Results                    55 ms   react-scripts test --testNamePattern=CompleteChip component test
      ✔ CompleteChip.test.js            55 ms
        ✔ CompleteChip component test   55 ms
          ✔ no task, not visible        11 ms
          ✔ incomplete task, not visible  2 ms
          ✔ complete task, shows date   42 ms
```

Figure 10.2 – A screenshot of IntelliJ's 'CompleteChip component test' test execution results

Now that we know how to run the tests from the IntelliJ IDE, let us continue implementing the tests for the application's routes.

Testing the router

In the *Adding routing* section in *Chapter 7, Bootstrapping the React Project*, we included React Router in our project to provide a routing solution for our application. Then, in the *Adding the task pages to the application router* section in *Chapter 9, Creating the Main Application*, we configured the definitive routes for the task manager. Just like any of the application's features, these routes should be tested to make sure they follow the specifications and that they don't break in the future.

To be able to properly test the application routes, we'll need to be able to render complete application pages that require a Redux store configuration and an MUI theme provider. Let us create some utilities that will allow us to provide these settings in our tests.

Testing helpers

To host the testing helper files in our project, we'll create a new `src/__tests__` directory that will allow us to clearly distinguish this code from the production code. Next, we'll create a new `react-redux.js` file under this directory. You can find the complete source code for this file in the GitHub repository, but let us analyze the most relevant parts as follows:

```
import {render as testRender} from '@testing-library/
  react';

export const render = (ui, options = {}) => {
  const Wrapper = ({children}) =>
    <Provider store={store}><ThemeProvider theme=
      {theme}>{children}</ThemeProvider></Provider>;
  return testRender(ui, {wrapper: Wrapper, ...options});
};
```

The main purpose of this file is to expose a `render` function with the same signature as the React Testing Library's `render` function but wrapping the provided component (the `ui` parameter) with a `Wrapper` component. This wrapper provides a Redux store and MUI theme configuration that some components may require to render. Let us now see this function in action by implementing the tests for the task manager frontend application's routes.

Testing the application's routes and navigation system

The routes for the application are defined in the `src/App.js` file. Let us edit its corresponding test file, `App.test.js`, and replace its now outdated content with the router test suite. You can find the complete source code for this file at `https://github.com/PacktPublishing/Full-Stack-Quarkus-and-React/blob/main/chapter-10/src/main/frontend/src/App.js`.

The following code snippet contains the code that will set up and tear down the MSW server:

```
let server;
beforeAll(() => {
  server = setupServer();
```

```
      server.listen();
    });
  afterAll(() => {
    server.close();
  });
```

The `beforeAll` block contains a function that Jest will execute a *single* time before any of the tests in the surrounding `describe` block is executed. We'll use this kind of setup whenever we want to configure an initial context or state that is common to every test and won't be altered by any of them. The first statement, `setupServer`, configures the MSW server. We could optionally pass a set of initial MSW mock handlers as the first argument of the function. The second statement, `server.listen()`, starts the MSW server, which will now be able to accept requests.

The `afterAll` block contains a function that Jest will execute a *single* time after all of the executed tests in the surrounding `describe` block have finalized their execution. The arrow function contains a single statement that closes the MSW server, which will no longer accept connections.

The following snippet contains the preparation code required for each of the tests:

```
beforeEach(() => {
  window.sessionStorage.clear();
  server.resetHandlers();
  server.use(rest.all('/api/*', (req, res, ctx) => res
    (ctx.status(404))));
});
```

The snippet contains a `beforeEach` statement with a function that Jest will run *each* time before any of the tests in the surrounding `describe` block are executed. What we do here is prepare the context for each of the tests by cleaning up any residue that previous tests might have left. We start by clearing `sessionStorage`, which will effectively remove any JWT that might have been persisted by one of the other tests. Then, we remove any configured mocks from the MSW server by running the server's `resetHandlers()` function.

In the last statement, `server.use()`, we prepare an MSW request handler that will catch any HTTP request to a path that starts with `/api/` and respond with a 404 Not Found HTTP status code. This is used as a generic catch-all handler that will take care of any request we have not specifically prepared for a test case and that is not relevant to the tested scenario. Let us now see how to verify the router behavior for both users who have an active session and users who don't.

Testing redirection for logged-out users

The following test verifies that whenever the users navigate to the protected /tasks application path without an active or valid session, they are redirected to the /login path and the login form is presented to them:

```
test('logged out user visiting /tasks, redirects to /login', ()
=> {
  // Given
  window.history.pushState({}, '', '/tasks');
  // When
  render(<App />);
  // Then
  expect(window.location.pathname).toEqual('/login');
  expect(screen.getByText(/Sign in/, {selector: 'h1'}))
    .toBeInTheDocument();
});
```

In the Given phase, we use the web browser's History API pushState function to programmatically navigate to the /tasks URL. In the When phase, we invoke our customized render function to render the router component (App) into document.body. In the assertions phase, we verify that the current browser location has changed to the /login path; this asserts that our Redux store effectively handled the missing session and redirected the user to the login page route.

The last expectation uses the screen.getByText function to verify that the Sign in text is visible on the page. This time, we pass additional options in the function's second parameter to further refine the query and make sure that the Sign in text is visible within an h1 element. From the business logic perspective, this assertion is making sure that the router renders the Login page whenever a user navigates or is redirected to the /login path. Let us now implement the same test but for logged-in users instead.

Testing redirection for logged-in users

The test in the following code snippet verifies that whenever the users navigate to the protected /tasks application path with a valid session, they are not redirected and the Tasks page is presented to them:

```
test('logged in user visiting /tasks, displays /tasks', async
() => {
  // Given
  server.use(rest.post('/api/v1/auth/login', (req, res,
    ctx) =>
```

```
    res(ctx.status(200), ctx.text('a-jwt')))));
  await store.dispatch(login({name: 'user', password:
    'password'}));
  window.history.pushState({}, '', '/tasks');
  // When
  render(<App />);
  // Then
  expect(window.location.pathname).toEqual('/tasks');
  expect(screen.getByText(/Task manager/, {selector:
    '.MuiTypography-h6'}))
    .toBeInTheDocument();
});
```

The tested scenario contains asynchronous logic that needs to be waited for; a session needs to be loaded into the application's store, which requires backend queries to be completed. Notice how the nested test function is defined using the `async` keyword, which converts the function body into a promise. Jest infers the asynchronous nature of the function and will configure the test executor to wait for the promise to complete.

In the test arranging phase, we start by mocking the response to any HTTP POST requests performed to the `/api/v1/auth/login` path to return a fake JWT. Then, we directly dispatch a `login` action to the application's store using the `await` operator to wait for the asynchronous action to complete. This action is processed by the auth API Redux slice we implemented in the *Managing the frontend session* section in *Chapter 8*, *Creating the Login Page*, which internally performs the HTTP request we just mocked. At this point, when the test is executed, the context is set so that the application contains a valid user session.

Next, we programmatically navigate to the `/tasks` path and render the application router, just like we did for the logged-out user version of the test. However, since the user is now logged-in, the `Then` phase of the test contains different expectations. We first verify that the user has not been redirected elsewhere by asserting that the current browser path is `/tasks`. Then, we check that the `Tasks` page has been rendered by verifying the title displayed in the application's top bar, which should only be visible for authenticated users.

The `App.test.js` file contains additional tests for route-specific behavior, which follow a similar pattern; please make sure to check those out too on the GitHub repository. Let us execute them by clicking on the play button near the `describe` block and clicking on the **Run 'Router tests'** menu entry.

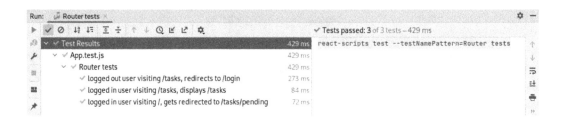

Figure 10.3 – A screenshot of IntelliJ's 'Router tests' test execution results

The tests should pass, assuring us that the application router is working as expected for the provided scenarios. These tests are exclusively focused on the routing layer by treating the router as the testing unit; however, each of the application routes can also be treated as part of the organic application's features. Let us now see how to write tests by treating each of the defined application features as the testing unit.

Testing the application's features

When implementing an application, especially if you're following a **test-driven development** (**TDD**) approach, it might be more interesting to write tests that verify the behavior of an actual application feature from the user's perspective rather than specific units of code from a component's technical perspective. The resulting tests will provide value from a business point of view and ensure, in an automated and sustainable way, that the application as a whole behaves according to what was requested in its specification. For this purpose, we'll create the following files:

- `src/auth/auth.test.js`
- `src/projects/projects.test.js`
- `src/tasks/tasks.test.js`
- `src/users/users.test.js`

Each of these files will contain tests that verify the application's features related to its containing module work as expected. For example, the `users.test.js` file contains tests that verify the task manager's user-related features, such as the ones in the following list, behave according to specification:

- `logged-in users can see their information`
- `logged-in users can change their password`
- `admins can navigate to the users page and see a list of users`

Notice how the names, rather than explaining the technical properties specific to a component (for example, `CompleteChip, with no task, is not visible`), reflect a feature of the application that adds value to the end users. Let us now analyze examples in more detail by focusing on the tests for the features related to the task manager's user authentication and authorization.

Testing the auth-related features

In *Chapter 8, Creating the Login Page*, we created the Login page to allow users to log into the application and implemented the auth API Redux slice to manage the user's session. In the auth.test.js file, we'll implement tests to verify that all of these features work according to what we specified. You can find the complete source code for this file at https://github.com/PacktPublishing/Full-Stack-Quarkus-and-React/blob/main/chapter-10/src/main/frontend/src/auth/auth.test.js. Let us analyze the most relevant parts and go over the concepts we haven't seen yet.

In the beforeEach code block, we can find the following statement:

```
server.use(rest.post('/api/v1/auth/login', (req, res, ctx)
  =>
      req.body.name === 'admin' && req.body.password ===
        'password' ?
      res(ctx.status(200), ctx.text('a-jwt')) :
          res(ctx.status(401))));
```

Since we are going to test the login form, most of the tests will perform an HTTP POST request to the /api/v1/auth/login endpoint. In this statement, we set an MSW handler to respond to these requests. If the request body contains the required fields, then the response will be successful and include a fake JWT. If the request doesn't contain valid credentials, a response with a 401 Unauthorized HTTP status code is emitted.

This test suite contains a nested describe block to group the tests related to the login feature as follows:

```
describe('login', () => {
  beforeEach(() => {
    window.history.pushState({}, '', '/login');
  });
```

Since every test in this group starts by loading the login page, we'll add this step to a common beforeEach Jest function.

The following test verifies the login feature's happy path, where a user logs into the application and is redirected to the pending tasks page, works as expected:

```
test('with valid credentials, user logs in', async () => {
  // Given
```

```
    render(<App />);
    // When
    userEvent.type(screen.getByLabelText(/Username/),
      'admin');
    userEvent.type(screen.getByLabelText(/Password/),
      'password');
    userEvent.click(screen.getByText(/Sign In/));
    // Then
    await waitFor(() => expect(window.location.pathname).
      toEqual('/tasks/pending'));
    expect(await screen.findByText(/Todo/, {selector:'h2'})).
      toBeInTheDocument();
});
```

In the When section, we're emulating a user typing some valid credentials in the applicable form fields and clicking on the sign in button. For this purpose, we use the following userEvent Testing Library functions to simulate user interactions with the provided DOM elements:

- userEvent.type

  This function fires the required DOM events to emulate what happens when the user types. The first parameter should target HTML input or textarea elements, and the second should contain the string representation of what the user is typing. In our case, we are using it to type the valid credentials to the username and password text fields. To retrieve the DOM nodes for these fields, we use the screen.getByLabelText function, which finds form elements with labels that match the provided regular expression in document.body.

- userEvent.click

  This function simulates a user clicking on the provided DOM element. In this case, we're simulating a user clicking on the Sign In button that should be visible in the login form.

In the Then section of the test, we wait for and assert that the user gets redirected to the /tasks/pending URL. We also make sure that the Tasks page is the one being rendered by checking that the page title (Todo) is present in the document. The screen function that we use in this case is findByText, instead of getByText, which does the same thing as getByText but waits asynchronously for queried elements to be on the page.

Notice how this test also verifies that the application router is working as expected. In the end, you need to balance whether there are corner cases that require specific tests for the router, such as those we provided in the *Testing the router* section, or whether some of these tests are no longer needed because their specification is covered by the feature test.

This file also contains tests to verify what happens when there are errors during the user's login workflow and the action doesn't complete. For example, `the login, with invalid credentials, user gets credentials error` test verifies that when users provide wrong credentials, they receive a `Snackbar` message informing them of the situation as follows:

```
expect(within(await screen.findByRole('alert')).getByText(/
Invalid credentials/)).toBeInTheDocument();
```

Let us now execute the tests on IntelliJ to make sure everything is working as expected. Just as we did before, we press the play button near the topmost `describe` block and click on the **Run 'auth module tests'** menu entry.

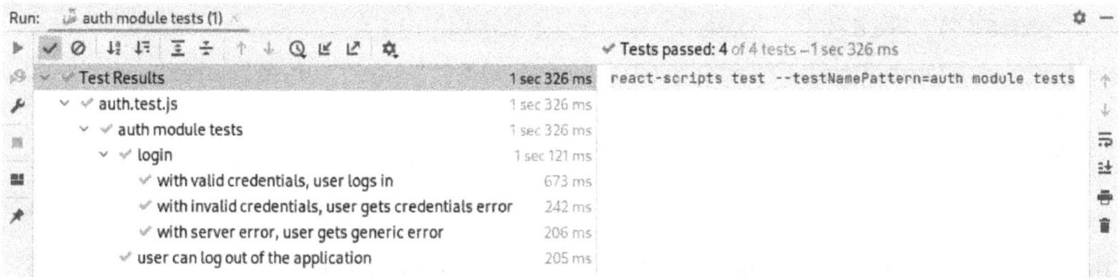

Figure 10.4 – A screenshot of IntelliJ's 'auth module tests' test execution results

We've now implemented enough tests to verify the auth features of the task manager. Let us continue by analyzing the tests that verify the task management-related features, which are the core of our application.

Testing the task management-related features

In *Chapter 9*, *Creating the Main Application*, we wrote the implementation for all of the features related to task management. In the `tasks.test.js` file, we'll implement tests to verify that all of these features work according to what we specified. You can find the complete source code for this file at `https://github.com/PacktPublishing/Full-Stack-Quarkus-and-React/blob/main/chapter-10/src/main/frontend/src/tasks/Tasks.js`. Let us analyze the most relevant parts and go over the concepts we haven't covered yet.

Dealing with tasks in the real application requires being able to choose from projects, having the logged-in user information, and so on. For this purpose, the global `beforeEach` block of this suite contains mocked responses for all of the HTTP resources, as you can see in the following snippet:

```
server.use(rest.get('/api/v1/tasks', (req, res, ctx) =>
  res(ctx.status(200), ctx.json([
    {id: 1, title: 'Pending task 1', description: 'A
```

```
        description', priority: 1, project: {id: 0, name:
          'Work stuff'}},
      {id: 2, title: 'Pending task 2', project: {id: 1, name:
        'Home stuff'}},
      {id: 3, title: 'Completed task 3', complete:
        '2015-10-21', project: {id: 1, name: 'Home stuff'}},
    ]))
  ));
server.use(rest.get('/api/v1/projects', (req, res, ctx) =>
  res(ctx.status(200), ctx.json([{id: 0, name: 'Work
    stuff'}, {id: 1, name: 'Home stuff'}]))));
server.use(rest.get('/api/v1/users/self', (req, res, ctx)
  =>
  res(ctx.status(200), ctx.json({id: 0, name: 'user',
    roles: ['user', 'admin']}))));
```

In this case, we're faking an initial list of tasks with different properties that will allow us to test the client-side filtering capabilities. In addition, we're mocking the list of projects from which the user will be able to choose and the information about the currently logged-in user. Since these are the only required endpoints, it's no longer needed for us to provide a generic catch-all handler as we did in the *Testing the application's routes and navigation system* section.

The test suite is organized into three different sections using Jest describe blocks as follows:

- users can create tasks

  This group of tests verifies the features related to new task creation. They are intended to ensure that the user can access the New Task dialog through the provided buttons in TopBar and on the Task lister page. They also ensure that the user can fill in the form data for the tasks and that this information is persisted when saved.

- users can list tasks

  These tests verify that the user can access the different task pages by clicking on their main drawer menu entry or directly navigating to the exposed route and that only the appropriate tasks are listed.

- users can edit tasks

  These tests verify that a user can click on a listed task to open the edit task dialog and that the edited tasks data is persisted in the backend when saved.

The test implementations are straightforward, readable, and self-explanatory. Let us execute them by clicking on the play button near the first `describe` block and clicking on the **Run 'tasks module tests'** menu entry.

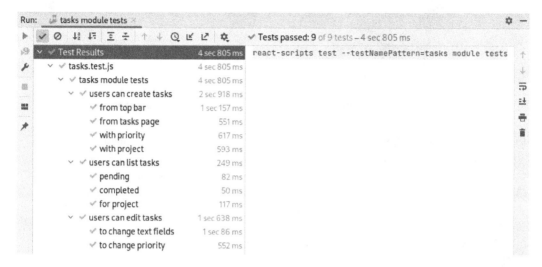

Figure 10.5 – A screenshot of IntelliJ's 'tasks module tests' test execution results

The feature tests for the rest of the modules, projects, and users follow the same patterns we described for the auth and task tests so we won't analyze them. Please check their implementation on the GitHub repository.

We've seen how to execute the tests from IntelliJ. Let us now learn how to run them from the command line.

Running the tests from the command line

So far, we've been running the tests from IntelliJ, which provides a convenient way to run a single test or a complete test suite. However, you might not be using IntelliJ, or even if you do, it's always important to know how to execute the tests using the command line. This is also the way you'd configure the test execution in a CI pipeline.

Our application was bootstrapped using Create React App, and one of its main features is the provision of scripts for your application that you don't need to maintain. This is the case for the test scripts too, which are linked in the `package.json` file as we can see in the following screenshot:

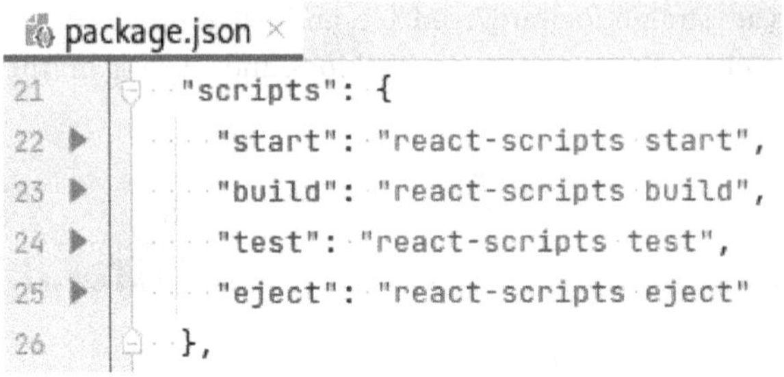

Figure 10.6 – A screenshot of the beginning of the scripts section in package.json

In the *Testing helpers* section, we added some testing helper files to a directory named __tests__. Unfortunately, Jest treats these files as tests too by default. We'll need to modify Jest's configuration so that the involved helper files are ignored. For this purpose, we'll add the following to the last part of the package.json file:

```
"jest": {
    "testMatch": ["**/?(*.)+(spec|test).[jt]s?(x)"]
}
```

With this testMatch property, we instruct Jest to search for tests that match the provided glob pattern, which will match all of our test files since they all have the .test.js suffix.

To run the tests from the command line, we just need to execute the following command from the frontend project root:

```
npm test
```

The default Create React App configuration will launch the tests in its interactive watch mode. This mode will run the tests that have changed since the last repository commit and also watch for changes to re-run those that are modified while the script is running. When running Jest in watch mode, your command should show the following messages:

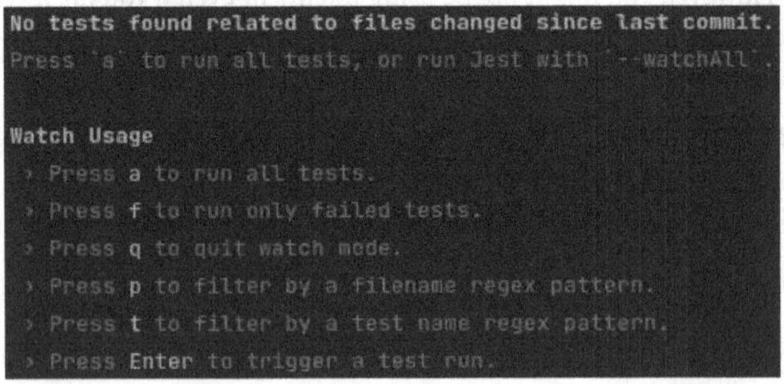

Figure 10.7 – A screenshot of the execution result of the npm test command

This mode can be useful in development mode and might improve your workflow performance. However, you might also want to run the complete test suite, not just those tests that have changed, in a non-interactive mode. You can accomplish this by executing the following command:

```
npm test -- --watchAll=false
```

The first - - is used to indicate that the subsequent arguments should be passed to the underlying `react-scripts test` command. The `--watchAll=false` flag is the Jest CLI configuration flag to disable watch mode. Once executed, you should see a test execution summary like the following:

```
Test Suites: 6 passed, 6 total
Tests:       24 passed, 24 total
Snapshots:   0 total
Time:        11.91 s
Ran all test suites.
```

Figure 10.8 – A screenshot of the execution result of the npm test -- --watchAll=false command

You might also be interested in finding the test coverage for the project. With the Create React App test script and Jest, it's as easy as specifying the `--coverage` argument, which can be used both in the watch and in the non-interactive modes. The following command will execute the tests and gather the coverage information with the watch mode disabled:

```
npm test -- --watchAll=false --coverage
```

The command should complete successfully and you should see a test execution summary like in *Figure 10.8* and a table with the detailed coverage results.

File	% Stmts	% Branch	% Funcs	% Lines	Uncovered Line #s
All files	87.71	73.14	84.12	87.5	
src	72.5	28.57	83.33	71.79	
App.js	100	100	100	100	
index.js	0	100	100	0	11-26
reportWebVitals.js	0	0	0	0	1-8
store.js	100	100	100	100	
useForm.js	100	0	100	100	3

Figure 10.9 – A screenshot of the execution result of the npm test -- --watchAll=false --coverage command

We have now completed the implementation of the required tests to ensure that the frontend side of the task manager behaves as expected. Our application is now much more reliable, and we should feel safe when we need to refactor or make changes to the project in the future.

Summary

In this chapter, we learned how to write tests for our JavaScript frontend application based on React, React Router, and Redux. We also provided complete test coverage for the task manager's features by implementing a complete portfolio of unit and integration tests. We started by learning about the required and recommended dependencies to implement our tests and added the missing ones. Then, we developed tests for our React components and our React Router configuration. We also wrote tests to verify the application's features from a user perspective and learned how to run them from the command line.

You should now be able to implement tests for your JavaScript frontend applications based on React, React Router, and Redux Toolkit using Jest, MSW, and React Testing Library. In the following chapter, we'll integrate the frontend and backend projects and create an API gateway in Quarkus so that the frontend application can be served from the backend.

Questions

1. What's the purpose of Jest's `describe` function?
2. How many times will the `beforeAll` function execute when running a complete test suite?
3. Can you run asynchronous tests in Jest?
4. What function would you use to emulate a user typing into a text field?
5. How can you calculate the code coverage of your tests?

11

Quarkus Integration

In this chapter, we'll learn how to integrate the React frontend and Quarkus backend projects so that the application can be deployed and distributed as a single service monolith. We'll start by analyzing the advantages and disadvantages of a microservice distributed architecture over a monolithic one, and evaluate why we should prefer a monolithic deployment for our task manager. Then, we'll learn how to set the Quarkus Maven configuration to build the frontend project and account for its generated resources. Next, we'll implement a Quarkus HTTP resource that will serve the frontend files and learn how to configure the application for a native build.

By the end of this chapter, you should be able to configure a Maven project and implement the required Quarkus logic to be able to serve a **single-page application** (**SPA**) from Quarkus. Serving a JavaScript frontend application from Quarkus can be very useful to ease deploying simple applications whose design allows them to be distributed as a single component.

We will be covering the following topics in this chapter:

- Distributing the application as a monolith versus a microservice
- Configuring the Quarkus application to build the frontend
- Creating an HTTP resource to serve the React application from Quarkus
- Configuring the native build

Technical requirements

You will need the latest Java JDK LTS version (at the time of writing, this is Java 17). In this book, we will be using Fedora Linux, but you can use Windows or macOS as well.

You will need the latest **Node.js** LTS version (at the time of writing, this is 16.15).

You will need a working Docker environment to deploy a PostgreSQL database and to create a Linux native image using Docker. There are Docker packages available for most Linux distributions. If you are on a Windows or macOS machine, you can install Docker Desktop.

You can download the full source code for this chapter from `https://github.com/PacktPublishing/Full-Stack-Quarkus-and-React/tree/main/chapter-11`.

Distributing the application as a monolith versus a microservice

In *Part 1, Creating a Backend with Quarkus*, we implemented an HTTP API and the core business logic for a task manager application. Then, in *Chapter 7, Bootstrapping the React Project*, we created the React application that acts as a frontend and main user interface for the task manager. With the currently decoupled project structure, it would be relatively easy to deploy the application in a distributed, microservice fashion, as separate components. Alternatively, we could apply some minor changes to integrate the frontend into the backend and distribute the task manager as a monolith. Let's learn what the advantages of exposing the application as separate microservices would be.

Advantages of a microservice architecture

Exposing the task manager as separate microservices would be quite easy with the current project structure. We'd need to package both the Java Quarkus-based backend and the Javascript React-based frontend into different container images and create a deployment configuration to integrate them. The following diagram shows what a distributed architecture for our current application could look like (logos attributions: The React logo is a trademark of Reactjs.org and is licensed under the Creative Commons. Attribution 4.0 International Public License. Quarkus® and the Quarkus logo are trademarks of Red Hat Inc and is licensed under the Apache License 2.0):

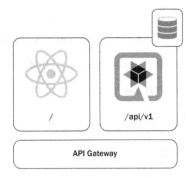

Figure 11.1 – A diagram of the task manager application distributed as separate microservices

The architecture would consist of four components – the frontend, the backend and its database, and an API gateway that would expose them. In this case, the API gateway would act as an intermediary between each component and the user, redirecting the traffic to the applicable service based on its target path. Requests to any path beginning with `/api/v1` would be routed to the Quarkus backend; every other request would be routed to the React frontend. In addition to routing and orchestrating the calls, the API gateway could also be configured to implement a **secure socket layer** (SSL).

It would be hard to find arguments in favor of deploying the application with this pattern since it's more complicated than deploying a single monolithic application. However, depending on how the application matures, it might make sense to reconsider this approach. The following diagram shows how the application could evolve in the future (logos attributions: The React logo is a trademark of Reactjs.org and is licensed under the Creative Commons. Attribution 4.0 International Public License. Quarkus® and the Quarkus logo are trademarks of Red Hat Inc and is licensed under the Apache License 2.0. Spring and the Spring logo are trademarks of Pivotal Software, Inc in the U.S. and other countries. KAFKA is a registered trademark of The Apache Software Foundation and has been licensed for use by Packt Publishing Ltd. Packt Publishing Ltd. has no affiliation with and is not endorsed by The Apache Software Foundation. The Keycloak logo is a trademark of Red Hat Inc and is licensed under the Apache License 2.0.):

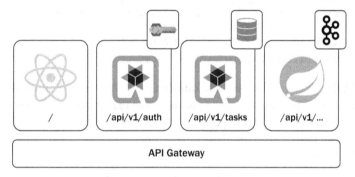

Figure 11.2 – A diagram of an evolved task manager application distributed as separate microservices

In this case, there are many more components. The React-based frontend service and the API gateway are still the same. However, there is now a specific Quarkus-based service exposed at the /api/v1/ auth path to handle everything related to authorization. This service relies on a KeyCloak server that should also be part of the deployment. There is also a specific Quarkus service for dealing with tasks exposed at the /api/v1/tasks path, which has its own database. Finally, there might be more services based on different technologies such as Spring, which could be connected to other services such as Kafka, a mail server, and so on.

In this more complex scenario, it would be easier to identify the advantages of a distributed architecture. Let's learn about some of them:

- **Easy horizontal scalability**: When one of the services runs out of resources, it's fairly easy to scale it just by providing more replicas or instances of that service. In addition, each service can be scaled independently. For example, the tasks microservice might need additional instances due to a high number of concurrent users, but the auth microservice might not because most of these users will only authenticate once. In a monolithic approach, you'd have to scale the complete application, which might consume unnecessary resources.

- **High redundancy for critical services**: Just like with scalability, with this architecture, you could easily provide additional instances of the most critical parts of your application. You could also put those critical service instances geographically closer to your users.

- **Easier deployment rollouts**: In a microservice architecture, each service should be completely decoupled from the rest. This means that services should be deployable independently. Rolling out a new version of a service shouldn't require a complete rollout of the application and all of its services. In turn, this means that the deployment process is so streamlined that it allows multiple deployments per day or even per hour.

- **Framework and programming language-agnostic**: *Figure 11.2* highlights this advantage, since some of the microservices are built with Quarkus and others with Spring. Having the application sliced into different services and components allows us to implement each of these components with different technologies. From a technical perspective, this is very good because it allows each component to be built with the technology or programming language that better suits the requirements. From a people management perspective, this is also good because you can have teams with different backgrounds and expertise working together.

These are just a few pros of a microservice-based architecture, but there are many more. However, all of these advantages come with a cost. Now, let's learn about some of the cons of this architectural pattern:

- **Additional infrastructure and automation required**: Microservices are very complex to operate. Instead of deploying a single application, you're running multiple services, in some cases hundreds or thousands. This requires specific and costly infrastructure such as Kubernetes, Amazon Web Services, and so on that can hold these complex distributions and help you automate the deployment and maintenance tasks.

- **Service-to-service communication issues**: Since the application is decoupled into multiple components or services that might require some sort of mutual communication, there is a higher chance of failure.

- **Transactionality and data consistency**: In a microservice or distributed architecture, each service relies on its own database or persistence mechanism. Managing transactions that span more than one service or extend to multiple domains, and achieving data consistency across them, can be very challenging and often requires complex patterns such as Sagas or **Command and Query Responsibility Segregation (CQRS)**.

- **Debugging and problem solving**: The increased complexity of the application architecture affects not only its IT operations but its maintenance and problem-solving tasks too. As a developer, debugging and solving a problem in a monolithic application can be hard already, but this difficulty grows exponentially when that same application is deployed as multiple isolated services.

These are just a few of the disadvantages of a microservice-based architecture, but there are many more. In the end, it's up to you to evaluate what's best for your project or application. You should carefully analyze the pros and cons of the microservice-based approach that's been applied to your project before considering its adoption. As we've learned, with our current project status, most of the advantages of a microservice-based architecture won't apply to our application. Now, let's see what a monolithic-based approach would look like.

The monolithic approach

We've seen that a microservice-based architecture has many advantages for some applications, especially those that are complex, consumed by a large number of users, and require big teams for their maintenance. However, right now, our application is really simple and won't benefit from most of the advantages of a distributed architecture, or at least not at the cost of the required additional complexity. The following diagram shows the components of a monolithic deployment for our application with just a few modifications (logos attributions: The React logo is a trademark of Reactjs.org and is licensed under the Creative Commons. Attribution 4.0 International Public License. Quarkus® and the Quarkus logo are trademarks of Red Hat Inc and is licensed under the Apache License 2.0):

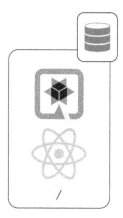

Figure 11.3 – A diagram of the task manager application's components distributed as a monolith

In this case, the application is distributed as two components – the application itself as an integrated version of the frontend and backend, and a database. Let's analyze why this is a good approach while considering some of the pros and cons we highlighted in the *Advantages of a microservice architecture* section:

- **No advanced scaling requirements**: Our application consists of two independent projects – the Quarkus backend and the React frontend. By distributing them as a single component, we lose the ability to scale each of them independently. However, we can still scale the complete monolith to provide high redundancy or to cope with higher user loads. As of now, the HTTP API exposed by the Quarkus application is only consumed by the React frontend. There are no mobile apps or third-party applications making use of this HTTP API either. Considering this fact, there would be no remarkable differences or requirements to scale just one of the components as opposed to scaling the complete monolith.

- **Small team**: The development and maintenance of the task manager could be managed by a very small team of developers. Splitting the application into several services would only increase the number of tasks and workload for this team.

- **Transactionality and data consistency**: When distributing the application as a monolith, you only require a single database or persistence provider. There aren't any complex transactions that affect different services. In this case, data consistency is handled by the database, which simplifies the implementation of workflows and transactions that span across multiple application domains.

- **No complex infrastructure needed**: Since there is no real need to split the application into multiple components, the tasks related to the monolithic deployment are much simpler. In this case, we'd just need a database (this could be a SaaS database or managed service) and the application deployment. This translates into cheaper infrastructure requirements and maintenance costs.

In general, you should always start with a well-designed monolith and continue evaluating whether migrating or evolving to a distributed architecture makes sense as time goes by. As we've learned, our application is very simple at the moment and wouldn't benefit from any of the microservice-based approach advantages. However, the application might evolve in the future and have more demanding requirements that could be leveraged by changing its architecture into a distributed one.

The current project structure would allow us to configure a distributed deployment without further modifications. However, it requires some minor changes to integrate the frontend with Quarkus so that the application acts as a single component. These changes can easily be rolled back in the future in case we reconsider splitting the application into multiple services.

Now, let's learn how we can configure the Quarkus build to build the React frontend too.

Configuring the Quarkus application to build the frontend

The first step of combining the React and Quarkus projects for a single deployment is configuring the Quarkus build process to run the frontend build and packaging tasks and account for the generated resources.

In the *Maven project (pom.xml)* section of *Chapter 1, Bootstrapping the Project*, we learned that the pom.xml file is the main unit of work for Maven and collects all the configuration details that will be used by Maven to build the project. Let's edit this file to add the required changes.

The following code snippet contains the changes in the pom.xml build section related to the required configuration so that Quarkus accounts for the resources generated from the frontend build process:

```
<resources>
  <resource>
    <directory>src/main/resources</directory>
  </resource>
  <resource>
    <directory>src/main/frontend/build</directory>
    <targetPath>frontend</targetPath>
```

```
    </resource>
  </resources>
```

This configuration affects the **Maven Resources Plugin,** which copies resources to the output or target directory. The first entry replicates the default setting for this plugin; all of the files located in the `src/main/resources` directory will be copied to the `target/classes` or `${project.build.outputDirectory}` directory. Despite this being the default setting, we need to specify this entry because we're overriding the overall resources configuration.

The second entry configures the plugin to copy all of the resources in the `src/main/frontend/build` directory to the `target/classes/frontend` directory. This configuration is the one responsible for allowing Quarkus to serve the frontend resources that will now be available on its classpath.

With the new resource configuration, we can already perform a Maven build that includes the frontend files. However, this would require manually executing npm commands on the frontend project to build the application before executing the Maven build. The next few code snippets contain a new Maven profile in our project's `pom.xml` file that will automatically invoke the required steps for the React build process. You can find the full source at `https://github.com/PacktPublishing/Full-Stack-Quarkus-and-React/blob/main/chapter-11/pom.xml`:

```
<profile>
  <id>frontend</id>
  <build>
    <plugins>
      <plugin>
        <groupId>org.codehaus.mojo</groupId>
        <artifactId>exec-maven-plugin</artifactId>
        <version>${exec-maven-plugin.version}</version>
```

In this snippet, we're defining the new `frontend` Maven profile whose main purpose is to run the frontend-related build tasks. There are several ways to accomplish this. In our case, we'll be leveraging **Exec Maven Plugin** since its configuration is straightforward and reflects the steps we want to perform through its `execution` configurations.

> **Exec Maven Plugin**
>
> Exec Maven Plugin is a plugin provided by the MojoHouse project. The plugin allows programs and Java programs to be executed in separate processes through its `exec` goal and the execution of Java programs in the same VM as the Maven build through its `java` goal.

Now, let's analyze the first `execution` configuration for the plugin:

```
<executions>
  <execution>
    <id>npm-install</id>
    <phase>generate-resources</phase>
    <goals>
      <goal>exec</goal>
    </goals>
    <configuration>
      <workingDirectory>src/main/frontend
        </workingDirectory>
      <executable>npm</executable>
      <arguments>
        <argument>install</argument>
      </arguments>
    </configuration>
  </execution>
```

The previous snippet contains the configuration for executing npm `install` to set up the frontend project. This execution binds the Exec Maven Plugin's exec goal to Maven's `generate-resources` build life cycle phase. This phase is executed by Maven before the resources are copied and the project compilation-related phases are performed. This guarantees that the resource directories we defined in our custom resource configuration will contain the required files before the Maven Resources Plugin copies them.

The `configuration` section contains settings for defining the working directory for the executed process – in this case, the relative path to the React project (`src/main/frontend`) – and the executable and arguments that will be executed by the plugin (npm `install`).

Now, let's study the second `execution` configuration for the new `frontend` Maven profile:

```
<execution>
  <id>npm-build</id>
  <phase>generate-resources</phase>
  <goals>
    <goal>exec</goal>
  </goals>
  <configuration>
    <workingDirectory>src/main/frontend
```

```
        </workingDirectory>
        <executable>npm</executable>
        <arguments>
          <argument>run</argument>
          <argument>build</argument>
        </arguments>
      </configuration>
    </execution>
  </executions>
```

This code snippet contains the configuration for the second execution, which will run npm run build to perform the build and file generation of the fronted project.

This execution has almost an identical configuration to the npm-install one but will execute npm run build instead. Note that the execution id is unique for each entry. Exec Maven Plugin will execute each entry with a unique id in the order defined in the pom.xml file. You should keep this in mind in case you want to add further execution configurations.

With that, we have configured the Maven build settings to account for the frontend files. However, we need a way to expose these files to the user. Let's see how we can achieve this by implementing a few HTTP endpoints in our Quarkus backend.

> **Quarkus Quinoa**
>
> Quarkus Quinoa is a Quarkus extension in the works (at the time of writing) that can help you automate, in an opinionated way, some of the tasks we've covered in this chapter. The extension is intended to ease both developing and serving SPAs, along with your Quarkus application. You can learn more about Quarkus Quinoa at https://quarkiverse.github.io/quarkiverse-docs/quarkus-quinoa/dev/index.html.

Creating an HTTP resource to serve the React application from Quarkus

One of the essential parts of the React and Quarkus integration is providing a way for the Quarkus backend to serve the React SPA frontend files. In *Figure 11.1*, we can see a diagram of the components of our application when deployed as microservices. One of the illustrated elements is the API gateway, which, in this case, is provided as an external component. For a monolithic approach, we'll need to implement an alternative to this gateway ourselves.

In the *Static resources* section of *Chapter 1*, *Bootstrapping the Project*, we learned that Quarkus automatically serves the static files in the `src/main/resources/META-INF` directory at the root path. You might be wondering why we should go through the trouble of implementing additional HTTP endpoints when we could simply configure the build process to copy the React build files to this directory or its effective compilation path (`target/classes/META-INF/resources`). This would work for simple applications that don't require frontend routing, but it won't work for ours. Let's proceed by implementing the additional HTTP endpoints by creating a new class in the `com.example.fullstack` package called `GatewayResource`:

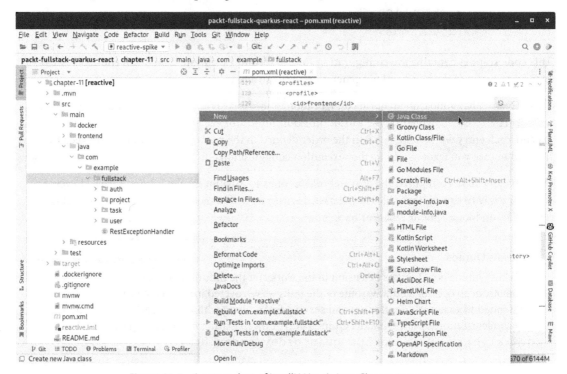

Figure 11.4 – A screenshot of IntelliJ New's Java Class menu entry

The purpose of this class is to expose an HTTP endpoint that can be consumed by the user's browser that will serve the frontend files in a way that is compatible with the frontend routing capabilities of our task manager. You can find the full source code for the `GatewayResource` class at `https://github.com/PacktPublishing/Full-Stack-Quarkus-and-React/blob/main/chapter-11/src/main/java/com/example/fullstack/GatewayResource.java`.

Now, let's analyze the most relevant parts:

```
@Path("/")
public class GatewayResource {
```

The class is annotated with the @Path annotation, which configures it to expose all of its defined endpoints at the root path. This configuration overrides the static resource's defaults. Quarkus will now serve the endpoints defined in this class instead of the content of the META-INF/resources directory.

The following snippet contains the code for getFrontendStaticFile, which implements the core logic for the gateway:

```
private static final String FALLBACK_RESOURCE =
  "/frontend/index.html";
@GET
@Path("/{fileName:.+}")
public RestResponse<InputStream> getFrontendStaticFile
(@PathParam("fileName") String fileName) throws IOException {
  final InputStream requestedFileStream = GatewayResource.
    class.getResourceAsStream("/frontend/" + fileName);
  final InputStream inputStream;
  inputStream = Objects.requireNonNullElseGet
    (requestedFileStream, () ->
    GatewayResource.class.getResourceAsStream
      (FALLBACK_RESOURCE));
  return RestResponse.ResponseBuilder
    .ok(inputStream)
    .cacheControl(CacheControl.valueOf("max-age=900"))
    .type(URLConnection.guessContentTypeFromStream
      (inputStream))
    .build();
}
```

The method is annotated with an @GET annotation so that it will only respond to HTTP GET requests. There's also an @Path annotation, which we've extensively covered in previous chapters. However, in this case, its value includes a fileName path parameter configured with a regular expression. This allows the endpoint to respond to HTTP requests that include forward slashes, such as /static/js/main.js.

> **Note**
>
> Even though `getFrontendStaticFile` has a very lenient regular expression matcher that matches most of the paths defined for other resources, such as `/users/1`, Quarkus will use this one as the last resort since it's the least specific. This is defined in the JAX-RS specification in the *3.7.2 Request matching* section, which has very strict sorting and precedence rules for matching URI expressions.

The method implementation starts by trying to find an application frontend resource with the specified filename or path. If a file is found, then it will be streamed back to the user. If none is found, then the `index.html` file will be streamed instead. This is the most important part since it's the one that'll enable frontend routing to work as expected. If a user visits the `/tasks/pending` or `/login` URL, there won't be any matching files, so the `index.html` file that contains the frontend application will be streamed instead. Once the browser loads it, the frontend router features will take care of the client-side redirection and load the appropriate page.

In the return statement, we build the response, including a cache configuration of 15 minutes, and the mime type for the file, which we compute by leveraging the `URLConnection.guessContentTypeFromStream` utility method.

The class also contains a `getFrontendRoot` method, which we can see in the following snippet:

```
@GET
@Path("/")
public RestResponse<InputStream> getFrontendRoot() throws
IOException {
   return getFrontendStaticFile("index.html");
}
```

This method is configured to reply to HTTP `GET` requests to the application root's path. The implementation calls the `getFrontendStaticFile` method with an `"index.html"` argument, which will stream the `/frontend/index.html` application resource back to the user.

With that, we've finished implementing the `GatewayResource` class, which will serve the frontend files from Quarkus, and the Maven build has been configured to include these files as application resources. We should now be able to run the application as a single component. Let's try it.

Running the application

In the previous chapters, we've always run the application in development mode by launching two processes: `./mvnw quarkus:dev` and `npm start`. With our recent changes, we should now be able to execute the application as a single service, just like we will when we deploy it in a production environment.

The only external component and requirement for our application is the database. If you don't have a PostgreSQL instance in your local environment, you can create one in a Docker container. Let's try this by running the following command:

```
docker run --rm --name postgresql -p 5432:5432 -e POSTGRES_
PASSWORD=pgpass -d postgres
```

This command starts a Docker container called `postgresql` and maps the container `5432` port (PostgreSQL default port) to the local `5432` port. The `--rm` flag instructs Docker to remove this container as soon as we stop it.

Now, let's build the application, including the frontend, by executing the following command:

```
./mvnw -Pfrontend clean package
```

The Maven build should start and we should be able to see log messages from the frontend build, as executed by Exec Maven Plugin:

Figure 11.5 – A screenshot of the Maven build logs showing the npm run build execution

The build should complete successfully and the complete application artifact files should be available in the `target/quarkus-app` directory.

Now, we can execute the application by running the following command:

```
java -Dquarkus.datasource.username=postgres -Dquarkus.
datasource.password=pgpass -Dquarkus.datasource.reactive.
url=postgresql://localhost:5432/postgres -Dquarkus.hibernate-
orm.database.generation=create -jar target/quarkus-app/quarkus-
run.jar
```

This command starts the application and provides the required configuration options for the data source. The credentials and the URL should match those of your PostgreSQL database. In this case, we're providing the credentials from the Docker container we launched previously. The application should start and we should be able to see the following log messages:

```
__ ____ __ _____ ___ __ ____ _____
--/ __ \/ / / / _ | / _ \/ //_/ / / / __/
-/ /_/ / /_/ / __ |/ , _/ ,< / /_/ /\ \
--_____/_/ |_/_/|_/_/|_|\____/___/
2022-09-16 13:17:31,774 INFO  [org.hib.rea.pro.imp.ReactiveIntegrator] (JPA Startup Thread: default-reactive) HR000001: Hibernate Reactive
2022-09-16 13:17:32,167 INFO  [io.quarkus] (main) reactive 1.0.0-SNAPSHOT on JVM (powered by Quarkus 2.10.2.Final) started in 1.453s. List
ening on: http://0.0.0.0:8080
2022-09-16 13:17:32,168 INFO  [io.quarkus] (main) Profile prod activated.
2022-09-16 13:17:32,168 INFO  [io.quarkus] (main) Installed features: [cdi, hibernate-orm, hibernate-reactive, hibernate-reactive-panache,
 reactive-pg-client, reactive-routes, resteasy-reactive, resteasy-reactive-jackson, security, smallrye-context-propagation, smallrye-jwt,
vertx]
```

Figure 11.6 – A screenshot of the application's execution log

Let's check the application in a browser by navigating to `http://localhost:8080`. Notice that we're now accessing the Quarkus application port, which is exposed at `8080`, instead of React's development server, which is exposed at `3000`:

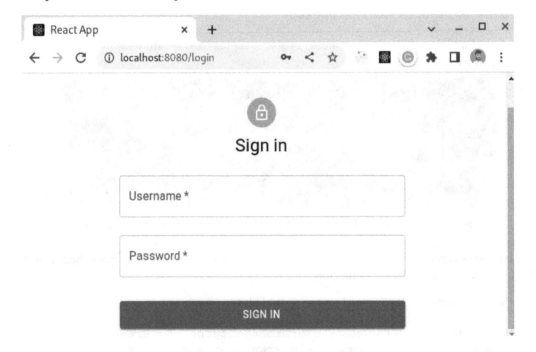

Figure 11.7 – A screenshot of a browser pointing to http://localhost:8080/login

The browser should automatically redirect us to `http://localhost:8080/login` and load the **Sign in** page. Since this is the production environment, only the `admin` user should be available. Let's log in to the application with the admin credentials (`admin/quarkus`) and play with the application. All of the implemented features should be working fine; we should be able to create projects, change our user's password, create tasks, mark them as complete, and so on. The following screenshot shows the application tasks page after playing with it for a while:

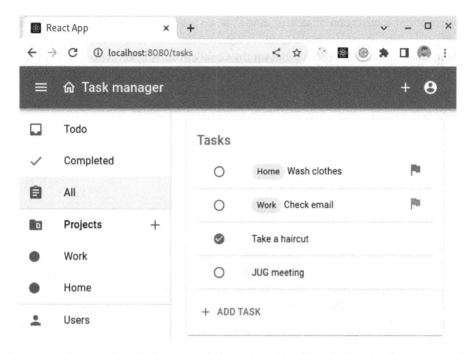

Figure 11.8 – A screenshot of a browser pointing to http://localhost:8080/tasks after some activity

The application is now fully functional and ready to be distributed as a monolith. However, if we want to package it as a native binary to improve its startup time and memory footprint, it won't work. Now, let's configure the application so that it can be distributed as a native image too.

Configuring the native build

In *Chapter 6*, *Building a Native Image*, we went over the necessary steps to package our backend Quarkus application into a native executable. With the recent changes we've made to include the frontend resources and the new `GatewayResource` class, we'll need to add further modifications to the native configuration so that everything works as expected. Let's start by configuring the native build so that it accounts for the frontend resources.

Including the frontend resources

In the *Including application resources* section of *Chapter 6*, *Building a Native Image*, we learned that the GraalVM native image uses AOT compilation and performs an aggressive static code analysis to find what resources, classes, methods, and so on should be included in the resulting binary image. We also learned that we can manually include additional resources by using the `quarkus.native.resources.includes` property. However, to include the frontend resources that the Maven Resource Plugin copies to the `src/target/classes/frontend` directory, we'll be using a different technique.

For our current purpose, relying on `quarkus.native.resources.includes` would be feasible too. However, we might encounter more complex scenarios in the future where this configuration property is insufficient. This approach relies on the GraalVM infrastructure instead of Quarkus' configuration. We'll start by creating a new `resources-config.json` file in the `src/main/resources` directory. The following code snippet contains the content for this file:

```
{
    "resources": [
        {
            "pattern": "^frontend.*"
        }
    ]
}
```

In this case, we're instructing GraalVM to include all of the resources in the paths matching the `^frontend.*` regular expression. This regular expression will match all of the frontend-generated files, including those in nested subdirectories. You can learn more about the resources configuration file and syntax in GraalVM's reference manual: `https://www.graalvm.org/22.1/reference-manual/native-image/Resources/`.

GraalVM won't detect the `resources-config.json` file automatically; we need to provide a configuration for this. Let's edit the `pom.xml` file and add the following property to the native profile:

```
<quarkus.native.additional-build-args>
    -H:ResourceConfigurationFiles=resources-config.json,
</quarkus.native.additional-build-args>
```

This setting configures Quarkus to add the `-H:ResourceConfigurationFiles=resources-config.json` flag to the GraalVM native image build arguments. If you want to include additional command-line arguments for the GraalVM `native-image` command, this would be the place to do so.

The resource configuration is now complete. However, we still need to add a minor modification to `GatewayResource` so that the task manager can be compiled to a fully working native image.

Fixing GatewayResource for native compilation

In general, Quarkus does an awesome job setting up and configuring GraalVM, and very little manual configuration is required from our side. However, sometimes, we might face situations where Quarkus misses some class or requires some extra tuning. For example, at the time of writing, there is a bug that prevents GraalVM from detecting the methods we implemented in the `GatewayResource` class.

In this case, to work around the bug, we'll be using the `@RegisterForReflection` annotation. Let's edit the `GatewayResource` class and annotate it, as shown in the following code snippet:

```
@Path("/")
@RegisterForReflection
public class GatewayResource {
```

The `@RegisterForReflection` annotation forces the annotated classes to be registered for reflection in the native image build process. So, despite this class not being correctly detected by GraalVM during the AOT compilation, by using this annotation, we force GraalVM to include it anyway.

With that, we've made the required changes to be able to compile a fully functional native binary for our application. We should now be able to package and run the application; let's try it.

Running the native application

To be able to run the application in native mode, we need to compile and package it first. We need to execute the Maven `clean package` goals, but in this case, we'll be using the `frontend` and `native` Maven profiles combined:

```
./mvnw -Pfrontend,native clean package
```

The build should complete successfully and the native application binary executable file should be available in `target/reactive-1.0.0-SNAPSHOT-runner`. We can now execute the application by running the following command:

```
./target/reactive-1.0.0-SNAPSHOT-runner -Dquarkus.datasource.
username=postgres -Dquarkus.datasource.password=pgpass
-Dquarkus.datasource.reactive.url=postgresql://localhost:5432/
postgres -Dquarkus.hibernate-orm.database.generation=create
```

The application should start and we should be able to see the following log messages:

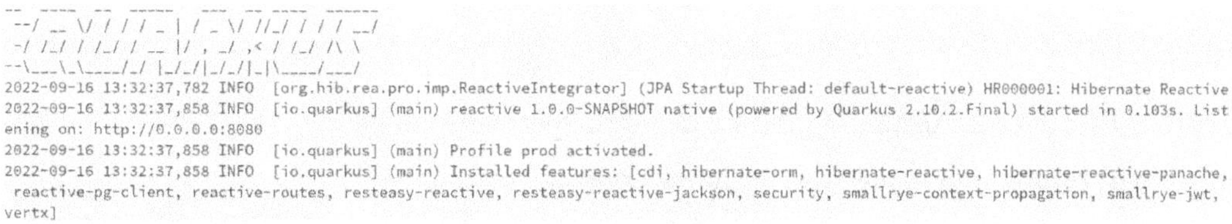

```
__  ____  __  _____   ___  __ ____  _____
 --/ __ \/ / / / _ | / _ \/ //_/ / / / __/
 -/ /_/ / /_/ / __ |/ , _/ ,< / /_/ /\ \
--_____/_/ |_/_/|_/_/|_|\____/___/
2022-09-16 13:32:37,782 INFO  [org.hib.rea.pro.imp.ReactiveIntegrator] (JPA Startup Thread: default-reactive) HR000001: Hibernate Reactive
2022-09-16 13:32:37,858 INFO  [io.quarkus] (main) reactive 1.0.0-SNAPSHOT native (powered by Quarkus 2.10.2.Final) started in 0.103s. List
ening on: http://0.0.0.0:8080
2022-09-16 13:32:37,858 INFO  [io.quarkus] (main) Profile prod activated.
2022-09-16 13:32:37,858 INFO  [io.quarkus] (main) Installed features: [cdi, hibernate-orm, hibernate-reactive, hibernate-reactive-panache,
 reactive-pg-client, reactive-routes, resteasy-reactive, resteasy-reactive-jackson, security, smallrye-context-propagation, smallrye-jwt,
vertx]
```

Figure 11.9 – A screenshot of the native application's execution log

> **Note**
>
> In the *Building the native image in a Docker Container* section of *Chapter 6, Building a Native Image*, we persisted a configuration to build the native image in a Linux Docker container. This means that the generated binary file will only be compatible with Linux. If you're on a different system, please skip this section.

Notice how the native startup time is now just a few milliseconds, which is orders of magnitude less than the JVM mode equivalent.

Just like for the JVM mode, we should be able to load the application by navigating to `http://localhost:8080` in our browser and log in using the `admin` user credentials. Every feature of the application should be fully functional.

> **Note**
>
> Remember to manually stop the PostgreSQL Docker container we've created to test run the application (`docker stop postgresql`) when you're finished. If this container is active while you're running or testing the application in dev mode, you'll face port conflict errors.

Summary

In this chapter, we learned how to integrate the React frontend and Quarkus backend projects so that the application can be deployed and distributed as a single component monolith. We started by learning the pros and cons of choosing a microservice architecture. Then, we evaluated why starting with a monolithic approach for our task manager made sense. Next, we learned how to configure the Maven build to include the frontend resources, and how to implement the required logic to serve them from Quarkus. Finally, we made the required changes to the native build configuration and learned how to run the application.

You should now be able to perform the required steps to be able to serve an SPA from Quarkus. In the next chapter, we'll learn how to deploy the application to Kubernetes.

Questions

1. What are three advantages of the microservice architecture?
2. What are three disadvantages of the microservice architecture?
3. How can we configure the Maven Resource Plugin so that it includes resources from multiple locations?
4. What are the main features of Exec Maven Plugin?
5. What are two ways to include additional resources in a native image?

Part 3– Deploying Your Application to the Cloud

This section focuses on the knowledge and skills required to create **continuous integration (CI)** pipelines for an application and its deployment. In this part, you will learn how to deploy an application to Kubernetes, in addition to Fly.io – a cloud **platform as a service (PaaS)** provider – and how to create CI pipelines using GitHub Actions.

In this part, we cover the following chapters:

- *Chapter 12, Deploying Your Application to Kubernetes*
- *Chapter 13, Deploying Your Application to Fly.io*
- *Chapter 14, Creating a Continuous Integration Pipeline*

12

Deploying Your Application to Kubernetes

In this chapter, we'll learn how to deploy our task manager application to Kubernetes. We'll start by learning about Kubernetes and containers and what the main features that have made these technologies so popular are. Then, we'll learn how to create a container image for our application and how to push it to an external container registry. Next, we'll create the required Kubernetes configuration files to be able to deploy our containerized application. Finally, we'll deploy our application to a minikube Kubernetes cluster.

By the end of this chapter, you should have a basic understanding of Kubernetes and application containers. You should also be able to create container images for your Quarkus applications and deploy them to Kubernetes clusters to make them publicly available.

We will be covering the following topics in this chapter:

- What is Kubernetes?
- Creating a container image
- Creating the cluster configuration manifests
- Deploying the task manager to Kubernetes

Technical requirements

You will need the latest Java JDK LTS version (Java 17 at the time of writing). In this book, we will be using Fedora Linux, but you can use Windows or macOS as well.

You will need the latest **Node.js** LTS version (16.15 at the time of writing).

You will need a working Docker environment to be able to run the Quarkus tests and to build and push container images using a Docker daemon. However, we'll analyze alternatives that won't require a Docker daemon. There are Docker packages available for most Linux distributions. If you are on a Windows or macOS machine, you can install Docker Desktop.

You can download the full source code for this chapter at `https://github.com/ PacktPublishing/Full-Stack-Quarkus-and-React/tree/main/chapter-12`.

What is Kubernetes?

Kubernetes is an open source solution for container orchestration, initially released by Google in 2014 and later transferred to the **Cloud Native Computing Foundation** (**CNCF**). It allows you to deploy, scale, and manage *container-based applications* in an automated, declarative way.

In the *Distributing the application as a monolith versus a microservice* section in *Chapter 11, Quarkus Integration*, we went through the advantages and disadvantages of distributing our application as a monolith or as multiple microservices. We learned that when dealing with a distributed architecture, it's advisable to rely on tools such as Kubernetes that help you automate and orchestrate the deployment and maintenance tasks for your services since there is a large number of them.

One of the main features of Kubernetes is that it abstracts away the underlying hardware infrastructure, providing a declarative interface to configure it. This allows you to define the requirements for the deployment of your application while Kubernetes takes care of setting up and configuring the hardware infrastructure for your needs. With this approach, developers can deploy applications by themselves just by providing declarative configurations for their workloads without the assistance of an operations team or a system administrator.

We won't get into too much detail about Kubernetes and its architecture since it's a vast and complex ecosystem. For now, it's fine if you understand that Kubernetes is a great option to orchestrate and automate the deployment of your applications and services that are packaged as *container images*.

In this book, we're building a simple task manager that can be deployed as a monolith that doesn't require a container orchestration solution such as Kubernetes. However, as we learned in *Chapter 11, Quarkus Integration*, our application might evolve in the future and increase in complexity, which might require switching to a distributed architecture that would benefit from a platform such as Kubernetes. In addition, we might also be subject to company-wide **non-functional requirements** (**NFRs**) that prescribe Kubernetes-based deployments. In this chapter, we'll learn how to leverage tools to simplify the complex process of deploying to Kubernetes, since it's very likely that, as developers, we'll end up dealing with this platform at some point. Let us continue by learning about containerized applications and their benefits.

What is a container-based application?

One of the biggest problems developers must deal with is the differences in the environments where their applications are executed. Surely we're all familiar with the phrase "It works on my machine." We might have even said it ourselves when we've experienced the frustration that our application or one of its components works perfectly fine in our development environment but doesn't in production. This is one of the reasons why Docker became so popular when it first released its Docker container solution in 2013.

Container images are a way to package and distribute your application along with all of its dependencies in an isolated, uniform, standard, lightweight, reproducible, and reliable way. For years, system administrators and operators have had to configure the production machines or **virtual machines** (**VMs**) to match the requirements where the applications had to be deployed. However, the introduction of containers and container images has allowed system administrators to treat applications as standard units that can be managed in a uniform and consistent way. Sysadmins no longer need to know the specific details and requirements of a given application since the container image encapsulates all of these details and allows the application to be deployed with the same conditions regardless of the environment they will run on.

The following list contains *some* of the many advantages of container-based applications:

- Environment consistency across each of the application phases (development, testing, and production)
- Lightweight runtimes
- More efficient hardware and resource usage
- Continuous development and continuous integration

These among many more advantages of container-based applications are the reasons why Kubernetes leverages and is built around this technology. Let us now learn how to set up a local Kubernetes cluster for our development needs using minikube.

Setting up a local cluster with minikube

Getting access to a production-grade Kubernetes cluster might be difficult or costly. Most of the Cloud providers such as Microsoft (AKS), Amazon (EKS), Google (GKE), and so on provide free trial periods to try their solutions; however, most of them require a credit card or an initial payment. Luckily, there are plenty of options to run a local Kubernetes cluster for development purposes. In this section, we'll see how to set up minikube, which is one of the most popular alternatives and is compatible with most operating systems.

Let us start by downloading minikube and adding it to our system's path. You can find detailed instructions for the mainstream operating systems at `https://minikube.sigs.k8s.io/docs/start/`.

If you followed the steps from the official minikube documentation (`https://minikube.sigs.k8s.io/docs/start/`), the `minikube` binary should now be available in your system's path. Let us start a cluster by invoking the following command in a console or terminal:

```
minikube start
```

minikube should start a cluster with the default settings for your system. If everything works OK, you should be able to see log messages similar to the ones in the following screenshot:

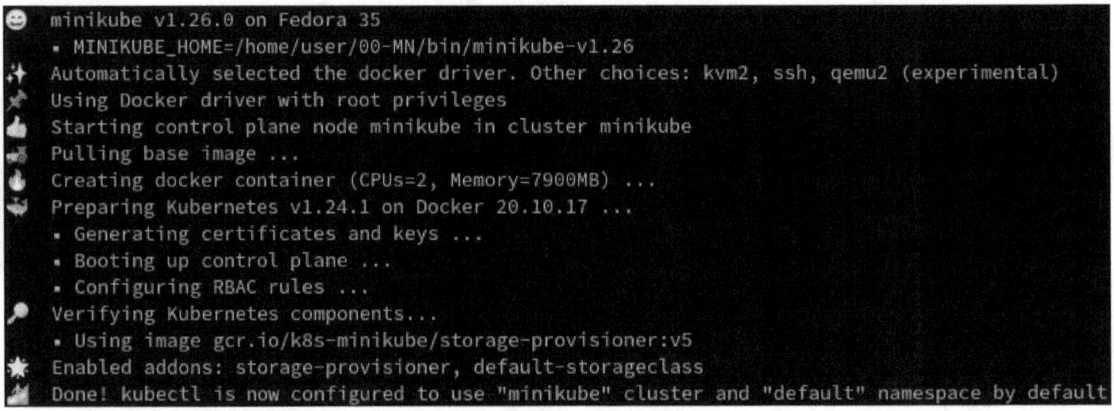

```
😄  minikube v1.26.0 on Fedora 35
    ▪ MINIKUBE_HOME=/home/user/00-MN/bin/minikube-v1.26
✨  Automatically selected the docker driver. Other choices: kvm2, ssh, qemu2 (experimental)
🐋  Using Docker driver with root privileges
👍  Starting control plane node minikube in cluster minikube
🚜  Pulling base image ...
🔥  Creating docker container (CPUs=2, Memory=7900MB) ...
🐳  Preparing Kubernetes v1.24.1 on Docker 20.10.17 ...
    ▪ Generating certificates and keys ...
    ▪ Booting up control plane ...
    ▪ Configuring RBAC rules ...
🔎  Verifying Kubernetes components...
    ▪ Using image gcr.io/k8s-minikube/storage-provisioner:v5
🌟  Enabled addons: storage-provisioner, default-storageclass
🏄  Done! kubectl is now configured to use "minikube" cluster and "default" namespace by default
```

Figure 12.1 – A screenshot of the minikube start command execution log

Your local Kubernetes cluster is now ready, however, we'll need to enable minikube's `ingress` addon so that we can complete all of the tasks in the chapter. Let us enable this addon by executing the following command:

```
minikube addons enable ingress
```

The command should complete successfully and you should be able to see the following messages:

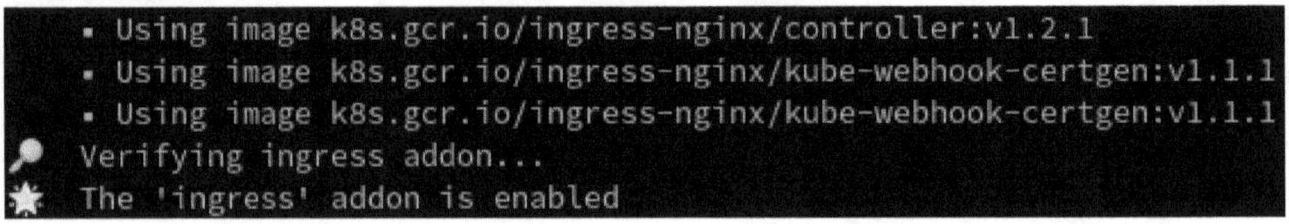

```
    ▪ Using image k8s.gcr.io/ingress-nginx/controller:v1.2.1
    ▪ Using image k8s.gcr.io/ingress-nginx/kube-webhook-certgen:v1.1.1
    ▪ Using image k8s.gcr.io/ingress-nginx/kube-webhook-certgen:v1.1.1
🔎  Verifying ingress addon...
🌟  The 'ingress' addon is enabled
```

Figure 12.2 – A screenshot of the minikube addons enable ingress command output

Our local development Kubernetes cluster is now ready to be used. Let us continue by learning how to build a container image for our task manager application.

Creating a container image

Regardless of your application's architecture, it's very likely that you need to distribute your application as a container image to be able to leverage any of the available cloud providers since container images are now the standard unit of distribution. As we've learned in the *What is a container-based application?* section, the fact that the operations teams can manage workloads consistently and uniformly by leveraging container technology is shifting part of their responsibilities to developers, who will now have to ship their applications packaged as containers.

In the Kubernetes world, *containers* and *container images* are the way to run your application. This means that one of the main requirements to be able to deploy your application is to package it up into one or more *container images* and push those images to an external registry available to your Kubernetes cluster. When deployed, Kubernetes will download the *container image* from a registry and run it in a *container* within a *Pod*.

> **Kubernetes Pod**
>
> A Kubernetes Pod is an object that represents the smallest deployable computing unit that can be managed in Kubernetes. Pods are the basic building block in Kubernetes. They contain a group of one or more application containers with shared network and storage resources. Whenever you need to run an application container in Kubernetes, regardless of the workload resource object you used to define it (**Deployment**, StatefulSet, Job, and so on), it will end up running within a Pod.

Being able to build container images of our applications and push them to external registries is a very valuable asset considering the new paradigm in systems administration and operations. Quarkus provides several extensions to build and push container images and to create the required configuration files to deploy them to Kubernetes. These extensions work great for some use cases; however, we'll be using **Eclipse JKube** and its **Kubernetes Maven Plugin** to build and deploy our application. The following are some of the reasons why we'll be using Eclipse JKube and its Kubernetes Maven Plugin:

- Support for multiple frameworks and libraries. Eclipse JKube works well for Quarkus but is compatible with other frameworks and libraries such as Spring Boot, Open Liberty, WildFly, and so on. Being able to reuse the same know-how for different use cases is very useful, especially if you deal with microservices implemented with different technologies.

- No need to deal with Dockerfiles. The most common way to define a container image is through a Dockerfile. Dockerfiles are just like a recipe or script that contains all of the required commands to assemble a container image. Quarkus provides its own set of Dockerfiles, which you can find in the `src/main/docker` directory. These files have a special syntax that you need to understand and require periodic maintenance tasks. With Eclipse JKube, there's no need to use Dockerfiles since it can automatically generate a container image for your application just by analyzing your project.

- Single dependency for all tasks and environments. Quarkus provides multiple extensions, both to generate container images and to generate and deploy cluster configuration manifests to Kubernetes. However, you'll need to mix and match several of these extensions to be able to achieve the same results as with Eclipse JKube. This becomes even harder when you have to deal with multiple target environments such as minikube, OpenShift, and so on. With Eclipse JKube, you only need to add the Kubernetes Maven Plugin to your project's plugin section.

- Additional development tools. Eclipse JKube can generate container images for your application, push them to an external registry, generate cluster configuration manifests, and generate and push Helm charts, among other features. It also includes developer-specific tools to improve the Kubernetes experience such as retrieving the logs for your application running in the cluster, opening a debug session, deploying and undeploying configuration manifests, and many more.

To be able to use our application's container image, we need to push it to an external image repository accessible to our cluster. There are plenty of alternatives available, both paid and free; you can even use minikube for this purpose. However, we'll be using Docker Hub since it's free and is publicly available to most Kubernetes clusters. Let us now see how to create an image repository to host the task manager application container image.

Creating an image repository in Docker Hub

If you don't have a Docker Hub account, you'll first have to sign up and create a new one by navigating to `https://hub.docker.com/`. There's a free personal plan that's suitable for our needs and just requires an email account.

Once you have your user account and are logged in to Docker Hub, it's time to create the image repository. We'll go to the repositories section (also available at `https://hub.docker.com/repositories`) and press the **Create Repository** button.

For the image name, we'll set the value `task-manager`, and you can also fill in the description if you want. Your form should look like the following screenshot once you've filled in all the fields:

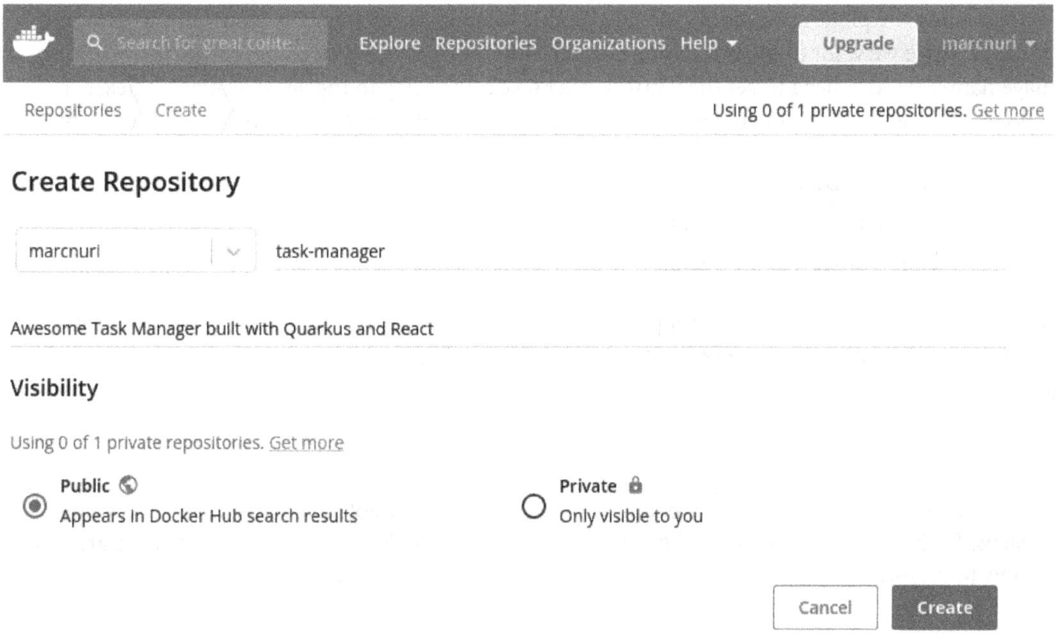

Figure 12.3 – A screenshot of the Docker Hub Create Repository user interface

We can now press the **Create** button and our repository should be ready. We have all of the requirements ready to be able to build and push our application's container image. Let us continue by configuring Eclipse JKube in our pom.xml file to perform the build.

Building a container image with Eclipse JKube

Building a container image with Eclipse JKube is as easy as adding the Kubernetes Maven Plugin to our project and providing some configuration flags. Let us start by opening the pom.xml file and adding the plugin to the <build><plugins> section as follows:

```
<plugin>
  <groupId>org.eclipse.jkube</groupId>
  <artifactId>kubernetes-maven-plugin</artifactId>
  <version>${jkube.version}</version>
</plugin>
```

Just by adding the plugin, we should already be able to build a container image for the application. However, since we want to push it to our Docker Hub repository, we need to change the resulting image name. We also need to set the `jkube.version` property to the latest available release. Let us do this by adding the following properties to the `<properties>` section in the `pom.xml` file:

```
<jkube.version>1.8.0</jkube.version>
<jkube.generator.name>marcnuri/task-manager:latest
  </jkube.generator.name>
```

The `jkube.generator.name` property can be used to configure the image name. In this case, we set it with the value we provided in Docker Hub like in *Figure 12.3* (note that you should replace `marcnuri` with your Docker Hub username instead).

> **Note**
>
> You will need an available Docker daemon in your system to be able to complete the following steps. In the *Using Jib to build and push a container image without Docker* section, we'll learn how to perform the same tasks without the need for the Docker daemon.

We can now proceed to build the container image for our application. However, we need to make sure that the application is built and packaged first. Let us do this by running the following command in a terminal:

```
./mvnw clean package -Pfrontend
```

The build should succeed and we should now be ready to create the container image by executing the following command in the same terminal:

```
./mvnw k8s:build
```

The `k8s:build` instruction is a Kubernetes Maven Plugin goal that instructs Eclipse JKube to build a container image for our application by analyzing the project and selecting the most suitable configuration values. The build should succeed and we should be able to see the following messages:

```
[INFO] --- kubernetes-maven-plugin:1.8.0:build (default-cli) @ reactive ---
[INFO] k8s: Running in Kubernetes mode
[INFO] k8s: Building Docker image in Kubernetes mode
[INFO] k8s: Running generator quarkus
[INFO] k8s: quarkus: Using Docker image quay.io/jkube/jkube-java:0.0.15 as base / builder
[INFO] k8s: [marcnuri/task-manager:latest] "quarkus": Created docker-build.tar in 340 milliseconds
[INFO] k8s: [marcnuri/task-manager:latest] "quarkus": Built image sha256:795f0
[INFO] k8s: [marcnuri/task-manager:latest] "quarkus": Tag with latest
[INFO] ------------------------------------------------------------------------
[INFO] BUILD SUCCESS
[INFO] ------------------------------------------------------------------------
```

Figure 12.4 – A screenshot of the result of the ./mvnw k8s:build command execution

Note that in this case, we've built the container image with the Quarkus application packaged as jar and not as a native binary. Creating a container image for the binary version of our application with Eclipse JKube would be as easy as adding the -Pnative Maven flag to the commands we executed in the previous code blocks (./mvnw clean package k8s:build -Pfrontend,native).

We have now built a container image for the task manager; however, it's only available in our local Docker registry. To be able to use it from our Kubernetes cluster, we need to push it to Docker Hub. Let us now learn how to achieve this using Eclipse JKube's Kubernetes Maven Plugin.

Pushing the container image to Docker Hub

Pushing an image to an external container image registry such as Docker Hub using Eclipse JKube's Kubernetes Maven Plugin is just as easy as building it. Since Docker Hub is a public registry, pushing to a repository is password-protected to prevent someone else from pushing an image to your repositories. Eclipse JKube has several ways to retrieve the required credentials to perform the push; the easiest is by reusing the ones stored in the local Docker configuration. In case you haven't done this yet, you'll need to log in to Docker Hub from the command line by running the following command:

```
docker login
```

You will be prompted to input your username and password for Docker Hub. After filling your credentials in, the login process should succeed and you should be able to see the following messages:

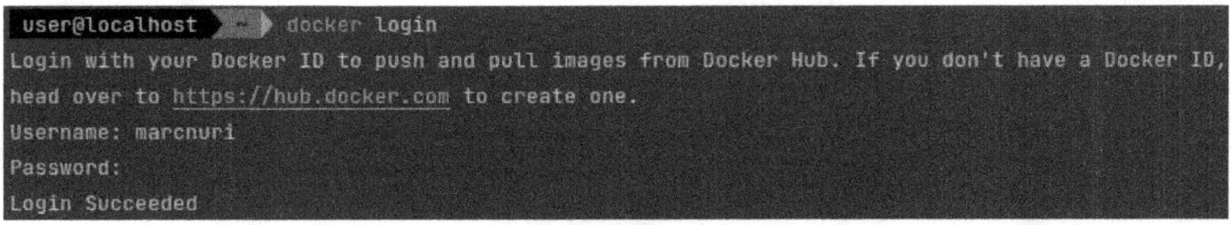

```
user@localhost  ─  docker login
Login with your Docker ID to push and pull images from Docker Hub. If you don't have a Docker ID,
head over to https://hub.docker.com to create one.
Username: marcnuri
Password:
Login Succeeded
```

Figure 12.5 – A screenshot of the result of the docker login command execution

We're now ready to push the image to Docker Hub. Let us do this by executing the following command in the same terminal we've been using until now:

```
./mvnw k8s:push
```

The k8s:push instruction is a Kubernetes Maven Plugin goal that instructs Eclipse JKube to push a container image that was previously built with the k8s:build goal. If everything works as expected, the Maven build should complete successfully and the image should be now available at Docker Hub. We can check this by visiting the URL of our repository; in our case, the URL is https://hub. docker.com/repository/docker/marcnuri/task-manager. An image tag should be available just like in the following screenshot:

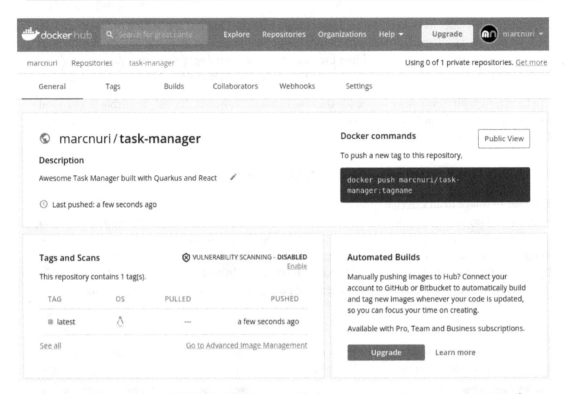

Figure 12.6 – A screenshot of the Docker Hub's marcnuri/task-manager repository management interface

Your image is now publicly available and ready to be consumed by any Kubernetes cluster with access to Docker Hub. However, we've relied on a local Docker daemon to perform the build and push tasks. There might be some environments, such as a **continuous integration** (**CI**) pipeline, where we don't have a Docker daemon available. Let us now see how to perform the same build and push tasks with Eclipse JKube's Kubernetes Maven Plugin but relying on **Jib** instead.

Using Jib to build and push a container image without Docker

Jib is an open source tool maintained by Google to build optimized container images for Java applications *without* a Docker daemon. One of the advantages of Eclipse JKube when compared to other alternatives is that you can switch the image build strategy just by providing a configuration flag.

Let us try to repeat the build and push process using Jib instead of Docker by running the Maven command with the following configuration flags:

```
./mvnw clean package k8s:build k8s:push -DskipTests -Djkube.
build.strategy=jib -Djkube.docker.push.username=marcnuri
-Djkube.docker.push.password=$password
```

Just by adding these three properties, Eclipse JKube's Kubernetes Maven Plugin performs the build using a completely different approach as follows:

- -DskipTests: This is a Maven setting to skip running the tests. Since our tests use Quarkus Dev Services that require a Docker daemon that might not be available, we specify this flag so we skip their invocation and possible failure.

- -Djkube.build.strategy=jib: This property instructs Eclipse JKube to use Jib instead of Docker, which is the default strategy. Eclipse JKube provides additional build strategies but they are mostly applicable to other Kubernetes cluster flavors such as OpenShift.

- -Djkube.docker.push.username=marcnuri: Since there is no Docker daemon available, there is likely no local Docker configuration from which Eclipse JKube can infer the credentials. This property is used to provide the Docker Hub username.

- -Djkube.docker.push.password=$password: The same as the username property, this one is used to configure the Docker Hub credentials password. In this case, we're using a password stored in the $password environment variable.

The Maven execution should complete successfully, and we should be able to see the following log messages with the build and push logs as performed by Jib:

```
[INFO] --- kubernetes-maven-plugin:1.8.0:push (default-cli) @ reactive ---
[INFO] k8s: Running in Kubernetes mode
[INFO] k8s: Building Docker image in Kubernetes mode
[INFO] k8s: Running generator quarkus
[INFO] k8s: quarkus: Using Docker image quay.io/jkube/jkube-java:0.0.15 as base / builder
[INFO] k8s: This push refers to: marcnuri/task-manager:latest
JIB> Containerizing application with the following files:
JIB> Container program arguments set to [/usr/local/s2i/run] (inherited from base image)
JIB> Building a single manifest
JIB> Checking existence of manifest for sha256:14f59cd78b33c864d3b4ae5b0757fb7f9c2f6eefcd19e2d268abd33251594510...
JIB> Skipping manifest existence check; system property set to false
JIB> Pushing manifest for latest...
JIB> [=========================== ] 98.9% complete > launching manifest list pushers
JIB> [============================] 100.0% complete
[INFO] --------------------------------------------------------------------
[INFO] BUILD SUCCESS
[INFO] --------------------------------------------------------------------
```

Figure 12.7 – A screenshot of the result of the k8s:push goal with Jib command execution

We are now able to build and push a container image of our application to an external container image registry even without Docker. Let us continue by learning how to generate the required configuration files to be able to deploy our application into Kubernetes.

Creating the cluster configuration manifests

Cluster configuration manifests are the files used by Kubernetes to be able to generate the required resources for your application. These are usually YAML files that contain a Kubernetes object declarative description that Kubernetes will interpret and try to satisfy by creating the requested objects and their resources using the underlying hardware infrastructure.

Writing and maintaining the cluster configuration files is a complex task that usually requires good knowledge and understanding of the different Kubernetes objects. Luckily for us, Eclipse JKube can also take care of creating these files for our application with a good set of opinionated defaults. Since our application has some special requirements such as a PostgreSQL database connection, we'll need to provide some minor configuration tweaks to override some of these default values.

Eclipse JKube provides different ways to override the default configuration, including setting some properties in the project pom.xml file or providing partial YAML files called fragments. Let us now learn how to modify the pom.xml file to tweak the cluster configuration manifests.

Adjusting the cluster manifests from the project's pom.xml

We'll start by editing the pom.xml file and changing the project's artifactId value to task-manager. The following snippet contains the affected change:

```
<artifactId>task-manager</artifactId>
```

Eclipse JKube uses the project's artifactId value by default to name the generated resources and objects, so it's advised that we change this to something that properly identifies our application. Next, we'll set some properties in the pom.xml file as follows:

```
<jkube.createExternalUrls>true</jkube.createExternalUrls>
<jkube.domain>192.168.49.2.nip.io</jkube.domain>
<postgresql.serviceName>postgresql
</postgresql.serviceName>
<jkube.enricher.jkube-project-label.group>
  ${project.artifactId}</jkube.enricher.
    jkube-project-label.group>
</properties>
```

Let's analyze what each of these properties does as follows:

- jkube.createExternalUrls: A property to configure Eclipse JKube to generate a Kubernetes **Ingress** object for our application to make it publicly available from a reachable URL.

- `jkube.domain`: A property that will be used as a host suffix for the generated Ingress. In this case, the resulting URL where our application will be available is `http://task-manager.192.168.49.2.nip.io`. You should replace the IP address in the property value (`192.168.49.2`) with the output of executing `minikube ip`, which reveals minikube's public IP address. *nip.io* is a free service that provides DNS services to expose any IP address, even private ones, as a hostname. You can learn more about this service at `https://nip.io`.

- `postgresql.serviceName`: This is not an Eclipse JKube configuration property. It will be used as a template variable to define the names of the Kubernetes objects we'll create to deploy a development PostgreSQL database.

- `jkube.enricher.jkube-project-label.group`: In Kubernetes, labels are key-value pairs attached to objects that are used to identify them in a way that is semantically relevant to users. Eclipse JKube automatically adds a few labels to the created objects, and one of them is `group`, which is usually computed from the project's `groupId`. With this property, we configure Eclipse JKube to add a label with a more specific value: `group: task-manager`.

Kubernetes Ingress

A Kubernetes Ingress is an object to configure external access and routing, usually HTTP, to a **Service** deployed in the cluster. Some other features that an Ingress provides are externally reachable URLs, name-based virtual hosting, SSL termination, load balancing, and more. Ingresses are not the only way to externally expose cluster services, but it's the most recommended approach.

Since our application relies on a PostgreSQL database, we'll need to provide additional configuration settings to pass the database credentials and URL information to the container application. Let's see how we can achieve this by taking advantage of Eclipse JKube fragments.

Adjusting the cluster manifests using fragments

Eclipse JKube automatically generates YAML cluster configuration files for your application by analyzing the project configuration and providing a set of opinionated defaults that should adjust to the most common use cases. However, each application has its nuances and specific requirements, so Eclipse JKube allows the tweaking of the generated YAML files by using fragments.

Fragments are *partial* definitions of Kubernetes objects that you can optionally provide and that Eclipse JKube will merge with the ones it generates for you. You can supply these fragments by creating the YAML files in the `src/main/jkube` directory of your project. Additionally, you can provide *complete* Kubernetes object definitions that will be deployed along with your application. These files should be created in the `src/main/jkube/raw` directory of your project.

Our task manager application requires a PostgreSQL database to work properly. When deployed to a production Kubernetes environment, the operations team will likely have already provisioned the database for us. However, since we're deploying the application to our local minikube cluster, we'll need to do this ourselves. We've created the required Kubernetes manifests for this purpose, and you can find them at `https://github.com/PacktPublishing/Full-Stack-Development-with-Quarkus-and-React/tree/main/chapter-12/src/main/jkube/raw`.

We won't go into the details of each of these YAML files since they are quite complex and fall out of the scope of this book. However, let's analyze closely the following `postgresql-secret.yml` file that defines a Kubernetes **Secret** since some of its properties will be reused by our customized fragment:

```
metadata:
  labels:
    app: ${postgresql.serviceName}
    group: ${project.artifactId}
  name: ${postgresql.serviceName}
stringData:
  username: quarkus-db
  password: the-s3cre7
```

The file contains the set of credentials (`username` and `password`) that will be used both by the PostgresSQL deployment to initialize the database and by our application to be able to log into the database service. In addition, the object definition contains a set of `labels` to properly identify the resource in the cluster. Note that the labels use placeholder variables to reference the properties we defined in the `pom.xml` file; Eclipse JKube will replace them when it generates the final cluster resources.

> **Kubernetes Secret**
>
> A Kubernetes Secret is an object used to store *confidential* data in key-value pairs. Pods can consume Secrets as environment variables, command-line arguments, or mounted as files in a volume. Note that Kubernetes Secrets, by default, store their data unencrypted.

To deploy our application into Kubernetes, Eclipse JKube will automatically generate a `Deployment` object with everything configured for us. However, if we deploy the application without further modifications, it won't be able to connect to the local PostgreSQL database.

> **Kubernetes Deployment**
>
> A Kubernetes Deployment is an object used to define workloads and applications in Kubernetes. It is a higher-level resource that runs multiple replicas of your application in identical Pods and takes care of replacing any instance that fails or might become unresponsive. Deployments ensure that your application has always at least the configured amount of replicas or instances available and ready to serve user requests.

In *Chapter 1*, *Bootstrapping the Project*, we learned that Quarkus relies on SmallRye and the Eclipse MicroProfile spec to deal with application properties. One of the alternatives to override these properties at runtime is by providing environment variables. We'll be using this technique to supply the database information to the Pod that runs our application's container. For this purpose, we'll create a new `deployment.yml` fragment file in the `src/main/jkube` directory. You can find the content for this file at `https://github.com/PacktPublishing/Full-Stack-Development-with-Quarkus-and-React/tree/main/chapter-12/src/main/jkube/deployment.yml`. Let's analyze the most relevant parts as follows:

```
containers:
  - env:
      - name: QUARKUS_DATASOURCE_USERNAME
        valueFrom:
          secretKeyRef:
            name: ${postgresql.serviceName}
            key: username
```

In the previous snippet, we're adding an environment variable to the first container defined in the `Deployment` resource. This variable will be initialized with a value that Kubernetes will read from a `Secret` matching the provided name: `${postgresql.serviceName}` (postgresql). This value will be read by Quarkus when it initializes the application, and will override the `quarkus.datasource.username` application property. The same pattern is applied to override the `quarkus.datasource.password` property.

In the following snippet, we override the data source URL so that Quarkus is able to connect to our local PostgreSQL service:

```
  - name: QUARKUS_DATASOURCE_REACTIVE_URL
    value: postgresql://${postgresql.serviceName}
      :5432/postgres
```

When Eclipse JKube processes this fragment, the `${postgresql.serviceName}` placeholder will be replaced by `postgresql`. This name matches the name of the Kubernetes Service we've defined in `src/main/jkube/raw/postgresql-svc.yml` and should be resolved by the Kubernetes DNS services. Kubernetes will route the traffic from this Service to the Pods it exposes.

Kubernetes Service

A Kubernetes Service is an object to expose an application running on a Pod or a set of Pods as a network service in the cluster. The recommended way to access a Pod's network from within the cluster is by exposing it through a Service. This way, clients don't need to know the location or address of an individual Pod. Once exposed, the application will be reachable by a DNS name that matches the Service name.

Since our database won't contain the schema for our application, we want it to be created automatically whenever the application starts. We will configure this with the following snippet:

```
- name: QUARKUS_HIBERNATE_ORM_DATABASE_GENERATION
  value: create
```

This environment variable overrides the `quarkus.hibernate-orm.database.generation` property, configuring it to perform only the database creation commands. This way, whenever the application starts or restarts, the database will be created only once in case it doesn't exist.

The last setting we'll override is the container image pull policy as follows:

```
imagePullPolicy: Always
```

The `imagePullPolicy` field configures how often Kubernetes should pull the container image from an external registry. Since we're still in the development phase, and we haven't provided a stable version for our container image, we'll set this value to `Always` to make sure that Kubernetes always deploys the freshest version of our application.

We have now completed all of the necessary adjustments in our project to be able to generate valid Kubernetes configuration manifests for our local minikube cluster. Let us now generate the cluster configuration manifests.

Creating the cluster configuration manifests with Eclipse JKube

Generating the cluster manifests with Eclipse JKube's Kubernetes Maven Plugin is as simple as executing a Maven command. Let's create them by running the following command in the same terminal session we've been using until now:

```
./mvnw k8s:resource
```

The Maven execution should complete successfully, and we should be able to see Eclipse JKube's log messages as follows:

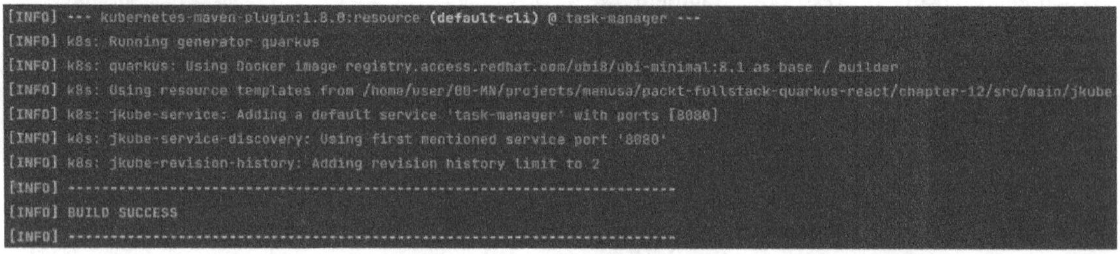

```
[INFO] --- kubernetes-maven-plugin:1.8.0:resource (default-cli) @ task-manager ---
[INFO] k8s: Running generator quarkus
[INFO] k8s: quarkus: Using Docker image registry.access.redhat.com/ubi8/ubi-minimal:8.1 as base / builder
[INFO] k8s: Using resource templates from /home/user/00-MN/projects/menusa/packt-fullstack-quarkus-react/chapter-12/src/main/jkube
[INFO] k8s: jkube-service: Adding a default service 'task-manager' with ports [8080]
[INFO] k8s: jkube-service-discovery: Using first mentioned service port '8080'
[INFO] k8s: jkube-revision-history: Adding revision history limit to 2
[INFO] ------------------------------------------------------------------------
[INFO] BUILD SUCCESS
[INFO] ------------------------------------------------------------------------
```

Figure 12.8 – A screenshot of the result of the Maven k8s:resource goal execution

The final cluster configuration YAML files generated and processed by Eclipse JKube should now be available in the `target/classes/META-INF/jkube` directory. You could now store them somewhere else, inspect them, and so on. However, if you are going to deploy them with Eclipse JKube, you don't need to worry about them. Let us now deploy the manifests using Eclipse JKube's Kubernetes Maven Plugin.

Deploying the task manager to minikube

Eclipse JKube's Kubernetes Maven Plugin includes another goal that allows you to deploy the application into Kubernetes from the generated YAML files without the need for other tools such as **kubectl**. Let's do this by executing the following command in the same terminal session we've been using so far:

```
./mvnw k8s:apply
```

The Maven execution should complete successfully, and we should be able to see Eclipse JKube's log messages related to the Kubernetes object creations as follows:

```
[INFO] --- kubernetes-maven-plugin:1.8.0:apply (default-cli) @ task-manager ---
[INFO] k8s: Using Kubernetes at https://192.168.49.2:8443/ in namespace null with manifest /home/user/06-MN/pro
jects/manusa/packt-fullstack-quarkus-react/chapter-12/target/classes/META-INF/jkube/kubernetes.yml
[INFO] k8s: Creating a Secret from kubernetes.yml namespace default name postgresql
[INFO] k8s: Created Secret: target/jkube/applyJson/default/secret-postgresql-2.json
[INFO] k8s: Creating a Service from kubernetes.yml namespace default name postgresql
[INFO] k8s: Created Service: target/jkube/applyJson/default/service-postgresql-2.json
[INFO] k8s: Creating a Service from kubernetes.yml namespace default name task-manager
[INFO] k8s: Created Service: target/jkube/applyJson/default/service-task-manager-2.json
[INFO] k8s: Applying PersistentVolume postgresql from kubernetes.yml
[INFO] k8s: Creating a PersistentVolumeClaim from kubernetes.yml namespace default name postgresql
[INFO] k8s: Created PersistentVolumeClaim: target/jkube/applyJson/default/persistentvolumeclaim-postgresql.json
[INFO] k8s: Creating a Deployment from kubernetes.yml namespace default name postgresql
[INFO] k8s: Created Deployment: target/jkube/applyJson/default/deployment-postgresql-2.json
[INFO] k8s: Creating a Deployment from kubernetes.yml namespace default name task-manager
[INFO] k8s: Created Deployment: target/jkube/applyJson/default/deployment-task-manager-2.json
[INFO] k8s: Applying Ingress task-manager from kubernetes.yml
[INFO] k8s: HINT: Use the command `kubectl get pods -w` to watch your pods start up
[INFO] ------------------------------------------------------------------------
[INFO] BUILD SUCCESS
[INFO] ------------------------------------------------------------------------
```

Figure 12.9 – A screenshot of the result of the Maven k8s:apply goal execution

The application should be ready and we should be able to navigate to the exposed URL, in our case `http://task-manager.192.168.49.2.nip.io`, and test our application as follows:

Figure 12.10 – A screenshot of a browser pointing to http://task-manager.192.168.49.2.nip.io/login

Once we've finished testing our application, we can remove its deployment and clean up our local cluster by leveraging another one of Eclipse JKube's Kubernetes Maven Plugin goals. Let us do this by executing the following command:

```
./mvnw k8s:undeploy
```

The command should complete successfully and our application and its related services should be completely removed from the cluster.

Summary

In this chapter, we learned how to deploy our task manager application to Kubernetes. We started by learning about Kubernetes, and container technologies in general, and what their main advantages are compared to more traditional deployment strategies. Then, we learned how to create a container image for our application and the required cluster configuration manifests to be able to deploy it to Kubernetes using Eclipse JKube. Finally, we deployed the application into a local minikube Kubernetes cluster.

You should now have a basic understanding of Kubernetes. You should be able to create and publish container images for your applications, and know how to deploy them to Kubernetes to expose them to the world. In the following chapter, we'll learn how to deploy the application to Fly.io, one of the most popular Cloud application platforms with one of its free pricing plans.

Questions

1. What kind of applications does Kubernetes manage?

2. What is a container-based application?

3. What are the main advantages of application containers?

4. What is a Pod?

5. What is an Ingress?

6. Does Jib require Docker?

13

Deploying Your Application to Fly.io

In this chapter, we'll learn how to deploy our task manager application to **Fly.io**, one of the most popular cloud **Platform-as-a-Service** (**PaaS**) providers, where we'll make it publicly available to the rest of the world. We'll start by learning about Fly.io and why it's a good choice for deploying our application. Then, we'll study how to configure and build our project to be able to deploy it to Fly.io. Finally, we'll learn how to deploy the application along with its required services, and we'll test that everything works as expected.

By the end of this chapter, you should be able to deploy applications to Fly.io and make them available to the rest of the world. Being able to deploy your application to multiple platforms will give you flexibility and the ability to choose from multiple options when it comes to publishing your applications on the internet.

We will be covering the following topics in this chapter:

- Introducing Fly.io
- Configuring the project for Fly.io
- Deploying the task manager

Technical requirements

You will need the latest Java JDK LTS version (at the time of writing, this is Java 17). In this book, we will be using Fedora Linux, but you can use Windows or macOS as well.

You will need the latest **Node.js** LTS version (at the time of writing, this is 16.15).

You will need a working Docker environment to create a container image. There are Docker packages available for most Linux distributions. If you are on a Windows or macOS machine, you can install Docker Desktop.

You can download the full source code for this chapter from `https://github.com/ PacktPublishing/Full-Stack-Quarkus-and-React/tree/main/chapter-13`.

Introducing Fly.io

In *Chapter 12, Deploying Your Application to Kubernetes*, we learned how to containerize our application and deploy it to Kubernetes. In this case, we leveraged a local Minikube Kubernetes cluster since getting access to a public Kubernetes cluster might be difficult or expensive for you. Knowing how to deploy the application to Kubernetes is a valuable asset considering it's becoming the standard cloud platform these days. However, unless you used a public cluster instead of Minikube, so far, your application can only be consumed from your local machine.

In this chapter, we'll learn how to deploy the application to Fly.io, a new cloud PaaS provider that is becoming very popular thanks to its free plans. The main reason for choosing this platform is that it offers a free plan suitable for our application's needs. By the end of this chapter, our application will be publicly exposed and accessible from anywhere in the world.

Fly.io is a platform focused on running full-stack applications and their dependent services close to their target users. The platform runs physical servers in many cities across the globe, which allows developers to deploy their workloads close to where their users are, improving the latency and performance of the applications. Fly.io has been around for a few years now, but it became more relevant in the middle of 2022 when Heroku decided to stop offering its free plans. Similar to Heroku, Fly.io allows you to deploy and scale container-based images very easily, and offers a very good developer experience, so it's a perfect and almost transparent replacement for Heroku.

> **Heroku**
>
> Heroku was originally founded in 2007 and publicly launched in 2009, years before containers and Docker were a thing. It first provided support for Ruby, but more programming languages were soon added. By then, it became a very popular choice for deploying applications, mostly due to its simplicity, developer friendliness, easy scalability, and its free offerings for hobbyists. With the rise of Kubernetes and other platforms, its popularity is stagnating, and by the end of 2022, Heroku no longer has a free offering.

Fly.io has a free-tier plan that, at the time of writing, will allow you to run a small application, such as our task manager, and a persistence service, such as PostgreSQL, for free. Their billing requirements to consume the free tier have changed during the past months, so you might be required to provide a credit card to be able to consume it, though no charges should be applied.

Now, let's continue by learning how to configure our project to be able to deploy the application to Fly.io.

Configuring the project for Fly.io

When it comes to deployment options, Fly.io is a flexible platform that offers multiple alternatives. There are integrations with several programming languages and frameworks, and quick start guides with detailed documentation on how to deploy applications based on each of these technologies. Most of these choices involve Fly.io performing a container image build from your application or project sources. In addition, there's also the possibility to deploy a pre-built container image that is publicly available in Docker Hub. This is the most appropriate choice for our application's requirements since it is based on multiple programming languages and none of the Fly.io suggested approaches would work for us.

In the *Pushing the container image to Docker Hub* section of *Chapter 12, Deploying Your Application to Kubernetes*, we pushed a container image of the task manager packaged for *JVM* mode to Docker Hub. You might be wondering, why not just reuse this image instead of building a new one? If you recall, when we created the cluster configuration manifests, we provided additional configuration to propagate the database credentials to the application container. Fly.io doesn't have an option to set custom *dynamic* runtime environment variables, so we need to create a modified image that allows us to read these credentials from the Fly.io-specific environment variables and secrets that are provided at runtime by the platform.

In addition, considering that the free plan of Fly.io is restricted in terms of resources, creating a container image of a *native* binary is better suited for this deployment. The lower application memory footprint and size will help keep the virtual machine resource consumption low.

Since we already configured *Eclipse JKube* in our project to create the container images, and to deploy the task manager into Kubernetes, we'll also use it to create a Fly.io-adapted container image for our application. For the sake of simplicity, and to experiment with additional features of the *Kubernetes Maven Plugin*, we'll configure the build to use an adapted Dockerfile.

Creating a Dockerfile for Fly.io

Eclipse JKube supports building and pushing container images based on Dockerfiles too. Let's start by creating a new file named `Dockerfile.fly` in the project's root directory. You can find the content for this file at `https://github.com/PacktPublishing/Full-Stack-Quarkus-and-React/blob/main/chapter-13/Dockerfile.fly`. Let's analyze the most relevant parts:

```
FROM quay.io/quarkus/quarkus-micro-image:1.0
```

This statement sets the base image for our container. `quarkus-micro-image` is an image maintained by Quarkus and is the most suitable to containerize and run native binary executables.

In the following snippet, we are configuring the assembly for our image by copying the generated native executable to a file named `application` in the `deployments` directory. We also change the mode of this file to make it executable:

```
COPY fly/target/*-runner /deployments/application
RUN chmod a+x /deployments/*
```

The last statement contains the CMD instruction. For this base image, we just need to provide the invocation of the executable, along with some system property flags, to configure the application's behavior:

```
CMD ./deployments/application \
 -Dquarkus.http.host=0.0.0.0 \
 -Dquarkus.datasource.reactive.url=$DATABASE_URL \
 -Dquarkus.hibernate-orm.database.generation=create
```

Let's take a closer look at these properties:

- `quarkus.http.host`: With this property, we instruct Quarkus to serve traffic from all network interfaces.

- `quarkus.datasource.reactive.url`: This property overrides the database connection URL. Fly.io will populate the `$DATABASE_URL` environment variable at runtime with the attached PostgreSQL database service URL.

- `quarkus.hibernate-orm.database.generation`: This property configures Quarkus to create the database schema upon its first connection to the database.

To improve the docker build performance, we need to limit the number of files that JKube will send to the Docker-compatible daemon for the assembly process. For this purpose, we need to create a new file called `.jkube-dockerinclude` in the project's root directory. The following code snippet shows its content:

```
target/*-runner
```

With this entry, we configure JKube to only account for files whose filenames end with the `-runner` suffix in the `target` directory. Now that we've created the required configuration files, let's continue by configuring the application's Maven project to build and push the Fly.io container image to Docker Hub.

Setting up a Maven profile for Fly.io

To deploy an application to Fly.io using a remote Docker container image approach, we'll need to create a new Maven profile with the required JKube configuration to account for the `Dockerfile.fly` file we just created. Let's open the project's `pom.xml` file and create a new profile with a `fly id`. You can find the updated content for this file at `https://github.com/PacktPublishing/Full-Stack-Development-with-Quarkus-and-React/tree/main/chapter-13/pom.xml`. Let's analyze the most relevant part:

```
<configuration>
  <images>
    <image>
      <name>marcnuri/task-manager:fly</name>
      <build>
        <dockerFile>${project.basedir}/Dockerfile.fly
          </dockerFile>
        <assembly>
          <name>fly</name>
        </assembly>
      </build>
    </image>
  </images>
</configuration>
```

The previous code snippet has two main purposes:

- **Configuring the image name**: We configure the resulting image name with the `marcnuri/task-manager:fly` value. In the *Pushing the container image to Docker Hub* section in *Chapter 12, Deploying Your Application to Kubernetes*, we created a Docker Hub `repository marcnuri/task-manager` and pushed the image to the `latest` tag, which is the one that gets pulled by default. Since we're publishing an image to be consumed specifically by Fly.io, we'll tag and push the image with `fly` instead.

- **Configuring JKube to use the** `Dockerfile.fly` **file**: The `<build><dockerFile>` configuration instructs JKube to use a `Dockerfile` instead of analyzing the project and automatically generating one for us.

> **Image name**
>
> Note that the image name should contain your Docker Hub username instead of `marcnuri`.

The project configuration and required files are ready. Now, let's continue by building the container image and pushing it to the Docker Hub registry by repeating the steps we followed in the *Pushing the container image to Docker Hub* section of *Chapter 12, Deploying Your Application to Kubernetes*.

We will start by logging into Docker Hub from the command line by running the following command:

```
docker login
```

Then, we will build and push the container image by executing the following Maven command from the project's root directory:

```
./mvnw -Pnative,frontend,fly clean package k8s:build k8s:push
```

This command is moderately more complex than other Maven commands we've executed so far. Let's examine it closely:

- `-Pnative,frontend,fly`: With this flag, we instruct Maven to use the `native`, `frontend`, and `fly` defined profiles. The `native` profile forces Quarkus to perform a GraalVM native build. The `frontend` profile is the one we defined in the *Configuring the Quarkus application to build the frontend* section of *Chapter 11, Quarkus Integration*. Finally, `fly` enables the Maven profile we just defined in this section.

- `clean package k8s:build k8s:push`: This is the list of Maven goals to execute. We'll clean up any residual files from previous executions, repackage the application using a native binary, and build and push the container image using the Kubernetes Maven Plugin goals.

The command should complete successfully and the image tag should now be available and visible in the Docker Hub dashboard:

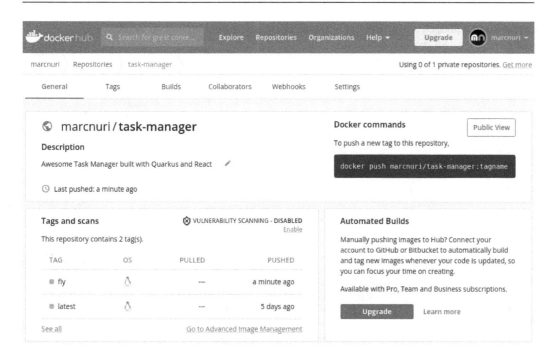

Figure 13.1 – A screenshot of the Docker Hub task manager repository user interface

Now, let's continue by creating a Fly.io account and publishing the application to the world by deploying the container image we just built and published.

Deploying the task manager

The main requirement to perform any operation in Fly.io is having its **command-line interface (CLI)** tool, `flyctl`, installed and available in your path. The installation process is straightforward and involves running just one or two commands in your operating system's terminal. You can find instructions for your specific platform at `https://fly.io/docs/hands-on/install-flyctl/`. Now, let's use this tool to create a new account or to log in to our existing account.

Creating a Fly.io account

There are several ways to create a Fly.io account. However, since we already have its CLI tool available, we'll use it for this purpose. The process is straightforward; the only requirement is having a valid email account. We can create a new account by running the following command:

```
flyctl auth signup
```

The command will open a new browser window where you should fill in your details:

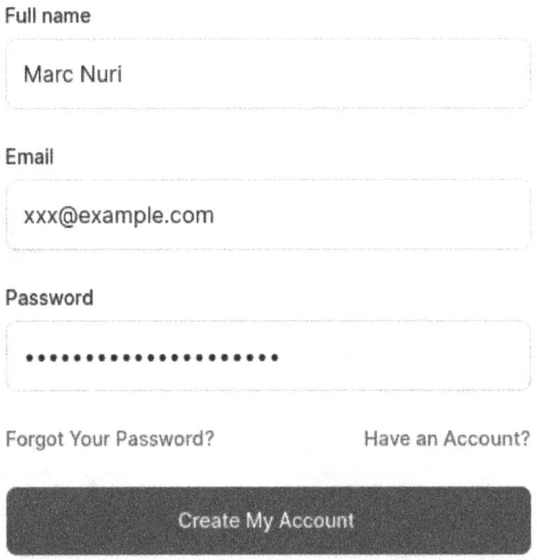

Full name

Marc Nuri

Email

xxx@example.com

Password

••••••••••••••••••••

Forgot Your Password?   Have an Account?

Create My Account

Figure 13.2 – A screenshot of Fly.io's create new account form

Once you've filled in your details and pressed the **Create My Account** button, the command execution should complete successfully and you should be logged in to your new account. You should also receive an email with a link to verify your account; you should complete this step before deploying the application.

If you already have a Fly.io account or the session you created with the new account expired, you can log into Fly.io with `flyctl`. Let's learn how.

Logging in to Fly.io

If you already have a Fly.io account, you should log in with your existing credentials. For that purpose, you need to run the following command:

```
flyctl auth login
```

The preceding command will open a new browser window where you should fill in your credentials:

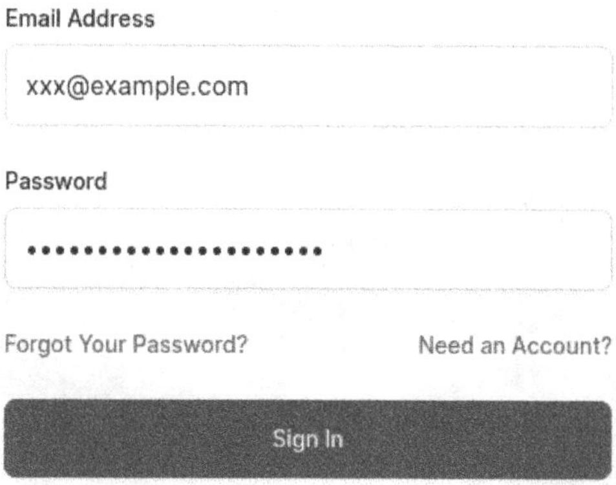

Email Address

xxx@example.com

Password

••••••••••••••••••

Forgot Your Password?   Need an Account?

Sign In

Figure 13.3 – A screenshot of the Fly.io login form

Once you've filled in your email and password and pressed the **Sign In** button, the command execution should complete successfully and you should be logged in to your account.

Creating a new application

The main requirement for deploying a workload to Fly.io is creating a new app. We can start this process by running the following command from the project root:

```
flyctl launch --image marcnuri/task-manager:fly
```

This command instructs `flyctl` to create a new application based on the `marcnuri/task-manager:fly` container image available from Docker Hub. You should replace the image name with the one applicable to your repository (`$dockerHubUserName/task-manager:fly`).

The process will start by prompting for an application name. This name must be unique for the entire Fly.io platform. In my case, I will use `task-manager`:

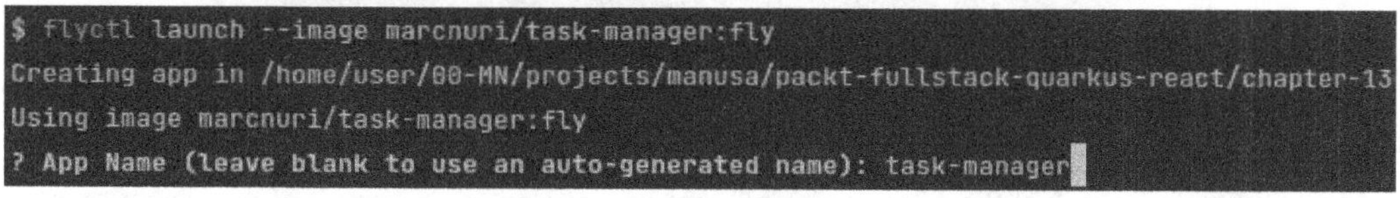

```
$ flyctl launch --image marcnuri/task-manager:fly
Creating app in /home/user/00-MN/projects/manusa/packt-fullstack-quarkus-react/chapter-13
Using image marcnuri/task-manager:fly
? App Name (leave blank to use an auto-generated name): task-manager
```

Figure 13.4 – A screenshot of flyctl launch prompting for an app name

Next, we need to select the region where the application will be deployed. One of the main features of Fly.io is that it can deploy your application to be close to where its users are. In my case, I'll select the `mad` region (Madrid).

You will also be prompted to optionally create a PostgreSQL database. Since our application relies on PostgreSQL, we'll confirm and create it by selecting the Development cluster which, at the time of writing, should be free:

```
? Select organization: Marc Nuri (personal)
? Select region: mad (Madrid, Spain)
Created app task-manager in organization personal
Wrote config file fly.toml
? Would you like to set up a Postgresql database now? Yes
For pricing information visit: https://fly.io/docs/about/pricing/#postgresql-clusters
? Select configuration: [Use arrows to move, type to filter]
> Development - Single node, 1x shared CPU, 256MB RAM, 1GB disk
  Production - Highly available, 1x shared CPU, 256MB RAM, 10GB disk
  Production - Highly available, 1x Dedicated CPU, 2GB RAM, 50GB disk
  Production - Highly available, 2x Dedicated CPU's, 4GB RAM, 100GB disk
  Specify custom configuration
```

Figure 13.5 – A screenshot of flyctl launch prompting you to set up a PostgreSQL service

After you've confirmed the desired configuration, `flyctl` should start the database deployment process. Once it's ready, the database credentials will be logged into the console. Unless you intend to connect manually to the database, you don't need them since they will be automatically injected into the `task-manager` application when deployed.

Finally, you will be asked whether you would like to deploy the application now. We'll answer no since, before deploying the application, it's convenient to go over the `fly.toml` file that the `flyctl launch` command execution has generated (we'll continue with the deployment phase later on in the *Deploying the app* section):

```
? Would you like to deploy now? No
Your app is ready. Deploy with `flyctl deploy`
```

Figure 13.6 – A screenshot of flyctl launch prompting to deploy the application

Now, let's take a closer look at the `fly.toml` file that was generated as a result of the execution of the `flyctl launch` command.

Analyzing the fly.toml app configuration

Fly.io uses the `fly.toml` file to configure an application's deployment. In our case, we used the `flyctl launch` command to automatically generate one for us. However, you can manually create and edit this file too. You can find the complete source of this file at `https://github.com/PacktPublishing/Full-Stack-Development-with-Quarkus-and-React/tree/main/chapter-13/fly.toml`. Let's go over the most relevant parts:

```
app = "task-manager"
```

The `app` name is the most important part of the configuration and is used to identify your application. It's also used to create the `fly.dev` hostname, which will make your application publicly available. The deployment of my application is available at `task-manager.fly.dev`.

The following snippet contains the `build` section of the configuration:

```
[build]
    image = "marcnuri/task-manager:fly"
```

We don't need Fly.io to build a container image for us since we already published a specially tuned image for this platform in Docker Hub. In our case, we just need to configure the `image` field with the name of our Fly.io-specific public image, which in my case is `marcnuri/task-manager:fly` (remember to replace this value with the name of your image). Fly.io will use this image instead of inferring our project configuration and building an opinionated container image.

The last important part of the configuration file is the services section, which contains the port mapping configuration for the application. The following snippet contains the relevant parts that affect the behavior of the task manager:

```
[[services]]
    internal_port = 8080
    [[services.ports]]
        force_https = true
        handlers = ["http"]
        port = 80
    [[services.ports]]
        handlers = ["tls", "http"]
        port = 443
```

The `internal_port` field is used to configure the port where the application will listen for connections. For Quarkus, `8080` is the default value, which is also the same value that `flyctl` uses as a default. If your application port doesn't match this value, you could either configure your application to listen on `8080` or modify the `internal_port` value.

The `service.ports` settings are used to publicly expose the internal port to the outside world. `flyctl` will automatically expose the application in the `80` (HTTP) and `443` (HTTPS) ports by adding two `service.ports` entries. Note that the port configuration for the `http` handler exposed at port 80 has a force_https = true that will automatically redirect the user to the tls handler and *enforce* an HTTPS connection.

Secure sockets layer/Transport layer security access

Secure sockets layer (**SSL**) and its successor, **Transport layer security** (**TLS**), are cryptographic protocols that provide communications security by establishing authenticated and encrypted links between networked computers. Although SSL has been deprecated for years, it is still common to refer to these technologies using the SSL or SSL/TLS acronyms. You should *always* access the application through a TLS-protected endpoint (HTTPS URL) to avoid exposing passwords or leaking any other sensitive information through non-encrypted standard HTTP connections.

We have now analyzed the most important settings of the fly.toml file and learned about their purpose. flyctl initialized the file with a few more settings that we didn't cover but they aren't relevant to our deployment. Of course, you are free to provide additional configurations that might be suitable for your application. You can learn more about these settings in the official Fly.io documentation at https://fly.io/docs/reference/configuration. Now, let's continue by deploying the application.

Deploying the app

The last step before our application becomes publicly available and accessible from the internet is deploying it to Fly.io. This process is as simple as running the following command from the project root directory:

```
flyctl deploy
```

This command should verify the application configuration, start the release process, and monitor its progress. After a few seconds, the execution should complete successfully and we should be able to see the following message logged to the console:

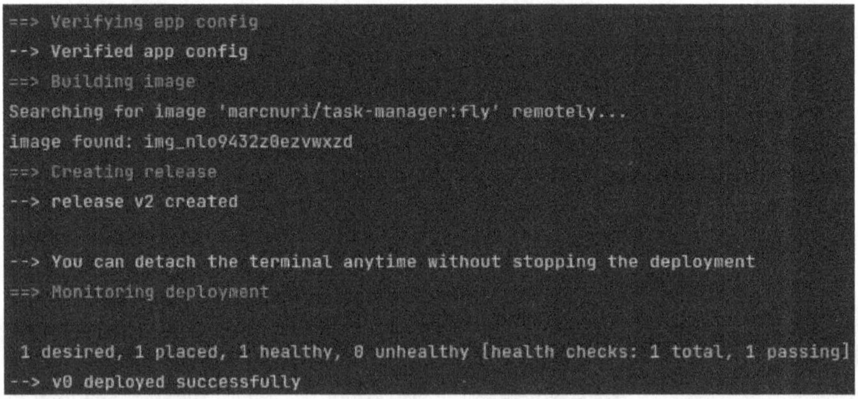

```
==> Verifying app config
--> Verified app config
==> Building image
Searching for image 'marcnuri/task-manager:fly' remotely...
image found: img_nlo9432z0ezvwxzd
==> Creating release
--> release v2 created

--> You can detach the terminal anytime without stopping the deployment
==> Monitoring deployment

 1 desired, 1 placed, 1 healthy, 0 unhealthy [health checks: 1 total, 1 passing]
--> v0 deployed successfully
```

Figure 13.7 – A screenshot of the result of executing flyctl deploy

You should also be able to see the deployment of the database and the application in the Fly.io dashboard for your organization. If you are using a personal organization, it will be available at `https://fly.io/dashboard/personal`:

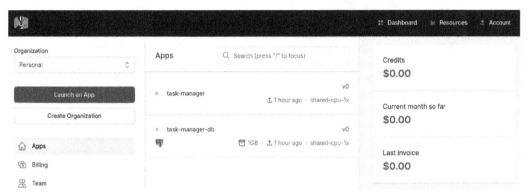

Figure 13.8 – A screenshot of the Fly.io dashboard showing the deployed application

Let's click on the `task-manager` (or the name of your app) entry to see the overview for this app:

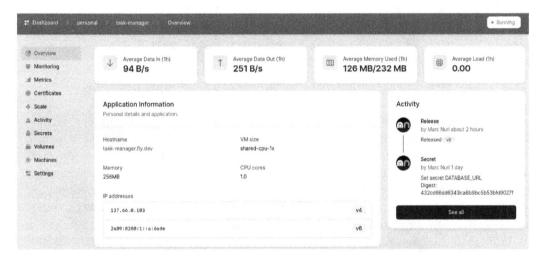

Figure 13.9 – A screenshot of the Fly.io application's overview interface

From this interface, we can see an overview of the most important details of our application. We can use it to check the application logs from the **Monitoring** tab, check the application health and metrics, scale the application, configure volumes, and so on.

The **Application Information** card shows the public URL for our application – in my case, this is `task-manager.fly.dev`. Let's click on this link to verify that our application is up and running and publicly available:

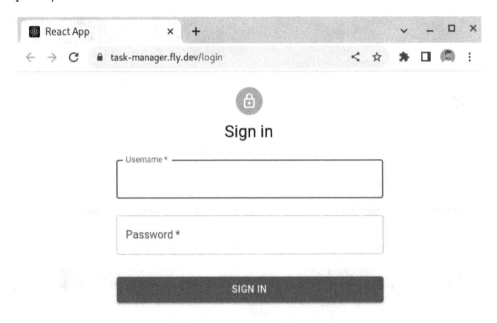

Figure 13.10 – A screenshot of the task manager deployed at Fly.io

The link should open a new tab and redirect you to the task manager login page (`https://task-manager.fly.dev/login`). You should be able to log into the application by providing the admin credentials (`admin/quarkus`) in the form and pressing the **SIGN IN** button. The application should let us in and we should be redirected to the task manager's landing page at `https://task-manager.fly.dev/tasks/pending`. We should now be able to use all of the application's features, such as creating new projects, creating new tasks, assigning priorities, setting tasks as completed, and so on. Most importantly, we should be able to play with the application from anywhere in the world with an internet connection. After a few minutes of using it, our application's dashboard should look more complete:

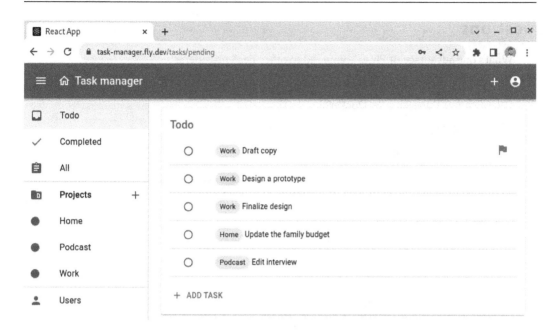

Figure 13.11 – A screenshot of the task manager after creating a few tasks and projects

With that, we have successfully deployed our application to Fly.io, but most importantly, we've made it publicly available to the world. After a few minutes of playing with the task manager, I'm sure that you're getting tons of ideas on how to improve it and add more features. The best part is that you can also ask for feedback from others since the application is online.

Summary

In this chapter, we learned how to deploy our task manager application to Fly.io. We started by learning about Fly.io, and why its free plan and its specific features make it a very good choice for deploying our application. Then, we learned how to configure our project to build and push a container image to Docker Hub specially tuned for Fly.io. Finally, we deployed the application and made sure everything worked as expected.

You should now be able to deploy applications to Fly.io and, most importantly, be able to share them with the rest of the world by exposing them on the internet. In the next chapter, we'll learn how to create a **continuous integration** (**CI**) pipeline for our application.

Questions

1. What's one advantage of Fly.io compared to other PaaS providers?
2. How can you override the configured database URL when running a Quarkus native application?
3. Does Eclipse JKube support Dockerfiles?
4. Where does `flyctl` read the app configuration from?
5. Can you deploy existing container images to Fly.io?

14

Creating a Continuous Integration Pipeline

In this chapter, we'll learn how to create a **continuous integration (CI)** pipeline with **GitHub Actions** for our task manager application. We'll start by learning about CI and why it's important in the context of GitHub Actions. Then, we'll learn how to create a GitHub repository to host our project and how to push our local files to the new repository. Finally, we'll learn how to create a GitHub Actions pipeline to build and test our project and understand how this fits in a CI and deployment software development practice.

By the end of this chapter, you should have a basic overview of GitHub Actions, and know how to implement CI workflows and pipelines for your projects. Being able to create CI pipelines for your project will allow you and your team to adopt agile software development practices while guaranteeing that your application doesn't break with each small change.

We will be covering the following topics in this chapter:

- Introducing GitHub Actions
- Creating and pushing the application into a GitHub repository
- Creating a GitHub Actions pipeline

Technical requirements

In this chapter, we'll create CI pipelines that run on the GitHub Actions server and **virtual machine (VM)** infrastructure. There is no need to have a Java JDK or Node.js setup.

You will need a Git setup on your machine to be able to persist your code and push it to GitHub. You will also need access to the internet and a GitHub account.

You can download the full source code for this chapter from `https://github.com/PacktPublishing/Full-Stack-Quarkus-and-React/tree/main/chapter-14`.

Introducing GitHub Actions

Before we dig deeper into GitHub and its Actions infrastructure, we need to have a better understanding of the CI, **continuous delivery**, and **continuous deployment** (**CD**) concepts (usually abbreviated as **CI/CD**), and how these *agile* practices can help in the software delivery process.

What is continuous integration?

The term *continuous integration* was originally coined by *Grady Booch* in his book *Object-Oriented Analysis and Design with Applications*, published in 1991. CI is a software development practice by which several developers commit their work on a single development project to a central repository where automated builds and tests are run to guarantee that these changes won't break the project. Further refinements of the CI term, especially those proposed by the **Extreme Programming** (**XP**) software development methodology, include repeating this operation multiple times a day.

CI tries to solve the problem where other development processes fail. In other methodologies, integrating several lines of work is a long and tedious process that often ends in unpredictable ways due to the bugs that might be introduced from the intersection of the different code changes. In CI, every change made to the project triggers a new build and a set of tests to ensure that the final product works. This means that every change is considered a standalone improvement and that these changes are integrated continuously, instead of delaying this process to a later time in the future, when the set of changes might have grown out of hand. Having a successful CI process requires a good infrastructure to be able to perform and run these builds and tests, as well as a good team culture where every member is committed to this practice. Continuous delivery and continuous deployment are built on top of CI but go a few steps further. Let's see what they involve.

What are continuous delivery and continuous deployment?

Continuous delivery is a natural extension of CI, where each set of changes is not only integrated into a central repository but is ready to be deployed into production after a successful build and test iteration, in a safe, quick, agile, and sustainable way. The main purpose of continuous delivery is to minimize the cost, time, and risk of delivering changes to a project or product.

In continuous delivery, the deployment phase to the end users or customers is triggered manually, and still involves a set of approvals. *Continuous deployment* accounts for this extra step and includes deploying and releasing the software to the end users within its fully automated pipeline. This practice allows teams to deliver new functionalities multiple times a day, drastically reducing the lead time compared to other approaches.

Besides the cultural change and commitment in the teams involved, the other main requirement to be able to adopt these practices is having a reliable platform where automated builds and tests are run. GitHub Actions is the CI/CD solution provided by GitHub. Let's learn a little bit more about its main features.

GitHub Actions overview

There are many tools available to help software development teams build their CI/CD pipelines. You can find self-hosted platforms such as Bamboo, Jenkins, TeamCity, and so on, as well as cloud-based **software-as-a-service (SaaS)** solutions such as CircleCI, GitLab CI/CD, Travis CI, and so on, most of which offer a free plan, especially for open source projects. In November 2019, GitHub made its own CI/CD SaaS solution, GitHub Actions, **generally available (GA)**, which has quickly become one of the most popular and widely extended choices.

GitHub Actions enables the automation of any software development workflow, allowing you to build, test, and deploy your project in the most suitable way for your team, just like most of the other solutions available. However, its tight integration with GitHub makes it the best candidate when building pipelines for GitHub-hosted projects. GitHub Actions is free for any public open source project and has (at the time of writing) a free 2,000-minute/month free quota for private projects.

One of the strongest points of GitHub Actions is its tight integration with the GitHub platform in general. Workflows can be triggered by responding to several events set off by a GitHub repository, such as pushing a commit to a given branch, creating a pull request, creating a tag or a release, and so on. The Actions **user interface (UI)** is also tightly integrated with the rest of the GitHub ecosystem. Execution reports are easily accessible from many of the repository's website sections. GitHub Actions has become an important part of the **Pull Request (PR)** user experience too, marking successful workflow executions as a requirement for a merge or being able to check the execution reports right from the PR interface are just a few examples of its tight integration.

GitHub Actions happen within a repository, so before creating a workflow for our task manager, we'll need to create a repository and push it to GitHub. In the next section, we'll learn how to use our GitHub account to create a GitHub repository and how to push our existing local project.

Creating and pushing the application into a GitHub repository

If you've been following along with this book, by now, you should have a directory in your filesystem that contains the complete application. If this is not the case, you can always download the ZIP file containing the application source code from `https://github.com/PacktPublishing/Full-Stack-Development-with-Quarkus-and-React/`, and use the code from `Chapter 13` as the starting point. In this section, we'll push the source code to a new GitHub repository.

> **GitHub user account**
>
> This section assumes you already have a GitHub user account. If this is not the case, you can easily create one by following the wizard at `https://github.com/signup`. The only requirement is having a valid email address.

Let's start by creating the new repository by clicking on the plus symbol and the **New repository** menu entry:

Figure 14.1 – A screenshot of GitHub's Create new pop-up menu

Now, let's fill in the details for the new repository. Here, you can introduce the repository name (in my case, `task-manager`) and leave the rest of the options with the default values:

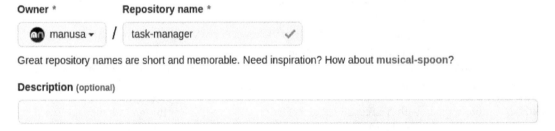

Figure 14.2 – A screenshot of the GitHub New repository form

GitHub should create the repository for you and redirect you to a page with instructions on how to initialize the local git repository. Before completing this step, we need to make sure that we can push to the remote repository either by setting up our SSH keys in our GitHub account or by creating a **personal access token** (**PAT**). A PAT is a type of token that can be used instead of a password to authenticate on GitHub; this is required when pushing git changes using HTTPS. Since pushing via HTTPS is the same procedure for all platforms, let's learn how to generate a PAT. Feel free to omit this step if you already have your local SSH keys set up for your user and you'll be using SSH instead.

Generating a PAT

You can generate a new PAT by visiting `https://github.com/settings/tokens` and clicking on the **Generate new token** button:

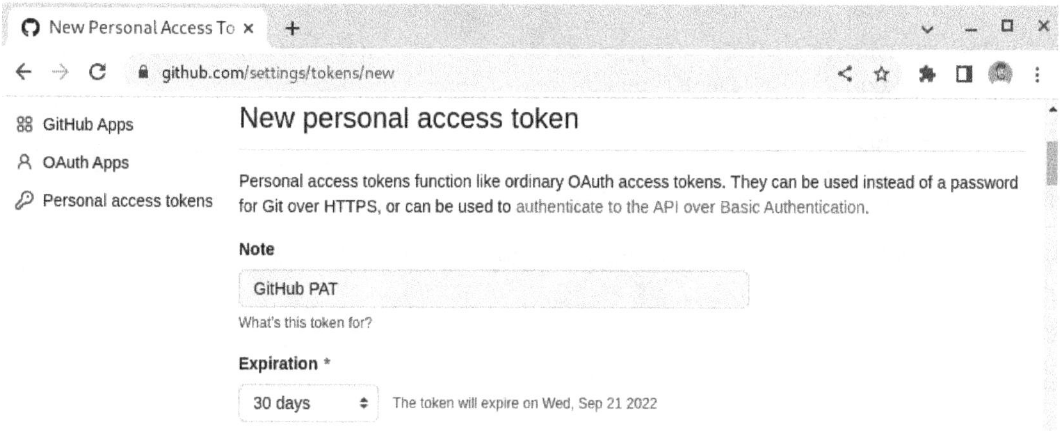

Figure 14.3 – A screenshot of the GitHub New personal access token form

In the wizard, make sure to select **repo** and **workflow**, and then click on the **Generate token** button to confirm. GitHub should generate the token for you and show it on the screen. Make sure to copy it and store it in a safe place since it won't be visible again. Let's continue by initializing the local git repository and pushing the project to GitHub.

Initializing the local git repository

We can initialize the local git repository by following these steps:

1. Open a Terminal on the project's root directory.

2. Initialize the repository by running the following command:

    ```
    git init -b main
    ```

3. Stage the project files in the current working tree and update the git index:

    ```
    git add .
    ```

4. Commit the staged changes:

    ```
    git commit -m 'Initial commit'
    ```

5. Configure the GitHub repository as the remote origin. Note that you should change the address with the one for your repo:

```
git remote add origin https://github.com/manusa/task-
manager.git
```

6. Push the branch to GitHub; when prompted, use your GitHub username and the **Personal Access Token (PAT)** instead of your account password:

```
git push -u origin main
```

If everything goes well, your repository should now contain the source code for the project:

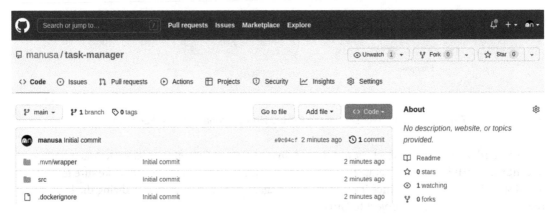

Figure 14.4 – A screenshot of the GitHub repository containing the project's files

Now that we have set up the local and remote repositories, we can create the pipeline to build and test the project. Let's continue by learning how to create the CI pipelines using GitHub Actions.

Creating a GitHub Actions pipeline

GitHub Actions pipelines or workflows are defined through YAML files that contain one or more jobs that are triggered by a set of specific git or GitHub events. The workflow YAML files must be located within the .github/workflows directory, so we'll start by creating this directory:

New Directory

.github/workflows|

Figure 14.5 – A screenshot of the IntelliJ New Directory dialog

Next, we can create the YAML file that will hold our workflow. For this purpose, we'll create a new file called `build-and-test.yaml` in the `.github/workflows` directory. You can find the full source code for the pipeline at `https://github.com/PacktPublishing/Full-Stack-Development-with-Quarkus-and-React/tree/main/chapter-14/.github/workflows/build-and-test.yaml`. Now, let's analyze the most relevant parts:

```
name: Build and Test
```

name will be used to identify the workflow when referenced from the GitHub repository UI. You should always provide a descriptive and meaningful name.

Next, we must define the events that will trigger a workflow run:

```
on:
  push:
    branches:
      - main
  pull_request:
```

Since this workflow will be used for CI validation, we want to trigger the pipeline whenever a commit is pushed to the `main` branch and whenever a user opens a pull request in the repo. These are just two types of events; there are many more. However, they highlight quite well how GitHub Actions can respond to git-specific events, as well as to broader GitHub events such as creating a pull request, an issue, a release, and so on. In addition, you can configure *manual* triggering by leveraging the `workflow_dispatch` event or *scheduled* invocations with the `scheduled` event configuration.

Each workflow can have one or more jobs that will, by default, run in parallel on different VMs. In the following snippet, we are defining a single job for our workflow:

```
jobs:
  build-and-test:
    name: Build and test
    runs-on: ubuntu-latest
```

Each job is defined as an entry within the `jobs` object. In our case, we defined a single entry with a `job_id` of `build-and-test` and a field called `name` with a more readable name for the job.

The `runs-on` field is used to define the type of machine that will be used to run the job. GitHub includes a set of runner machines for the major platforms (Ubuntu/Linux, macOS, and Windows Server); however, you can configure your own hosted runner too. In this case, we'll be using the latest Ubuntu version available, since we want to compile the application in Linux.

For each job, we need to define a sequence of `steps` or individual tasks to be executed in order. `steps` can either run a command, set up tasks, or use one of the predefined actions. In the following snippet, we are using one of the GitHub-provided actions to checkout the git repository:

```
steps:
  - name: Checkout
    uses: actions/checkout@v3
```

We must also use predefined actions to set up a Java and a NodeJS environment that's compatible with our project:

```
- name: Setup Node
  uses: actions/setup-node@v3
  with:
    node-version: 16
- name: Setup Java 17
  uses: actions/setup-java@v3
  with:
    java-version: 17
    distribution: temurin
```

In this case, the actions we are using are provided by GitHub, but you can use any of the ones provided in the GitHub Actions Marketplace (`https://github.com/marketplace?type=actions`) or even create and use your own.

For each of our `steps`, we set the `name` field with a meaningful description, and configure the `uses` field with the name and version of the GitHub action we want to use. Most actions accept some input parameters to further configure their behavior. In our case, we set the Node and Java versions to use, and the Java distribution provider.

The final part of the pipeline contains the actual build and test execution `steps`. In the following snippet, we are creating the step that will build the complete application, including the frontend and the backend, and that will run the Quarkus tests:

```
- name: Build and test
  run: ./mvnw -Pfrontend clean package
```

This step contains a field with a legible name, and the actual command to run. In this case, we are just reusing the command that we described in the *Running the application* section of *Chapter 11, Quarkus Integration.*

This command will install the npm dependencies and build the ReactJS frontend application; however, it won't run any of the frontend tests. We have two alternatives – create another Maven profile that runs the frontend tests when we package the application or add another step in the workflow to run them. Since, in this chapter, we're dealing with GitHub Actions, let's just add another step to the pipeline for this purpose:

```
- name: Test frontend
  run: cd src/main/frontend && npm test -- --all -
    watchAll=false
```

In this step, we are once again leveraging the `run` field to invoke the `npm` command to run the frontend tests. The `run` command will change the working directory to the frontend's project root, and then execute the command we described in the *Running the tests from the command line* section of *Chapter 10, Testing Your Frontend*.

The CI pipeline is ready to be tested – we just need to commit the changes and push them to our remote GitHub repository. From the same Terminal we used to initialize the git repository, let's run the following commands to commit and push our changes:

1.   Stage the GitHub Actions workflow file:

     ```
     git add .
     ```

2.   Commit the changes:

     ```
     git commit -m 'ci: build and test pipeline'
     ```

3.   Push the changes to GitHub. You will be prompted for your username and PAT:

     ```
     git push origin main
     ```

The commands should complete successfully and your workflow file should now be visible in your remote GitHub repository. Let's visit the **Actions** tab to see the workflow running:

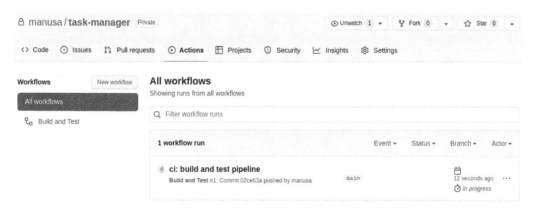

Figure 14.6 – A screenshot of the GitHub Actions dashboard for the project repository

The GitHub Actions dashboard should show us that the new **Build and test** workflow is running. The workflow run can be identified by the latest commit hash and message that was responsible for triggering the execution. We can also see that the workflow is running in the main branch. After a few minutes, the workflow should complete successfully. Let's click on it to see the execution summary:

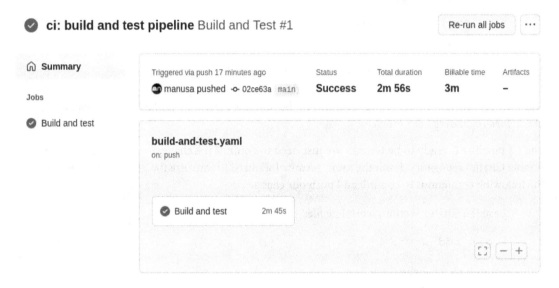

Figure 14.7 – A screenshot of the Build and test workflow run summary

Since our workflow is quite simple and only involves a single job, this view doesn't provide too much value. The summary becomes handier when there are more jobs involved and when some of the jobs depend on others. The following screenshot shows a summary of a more complex workflow from a different repository:

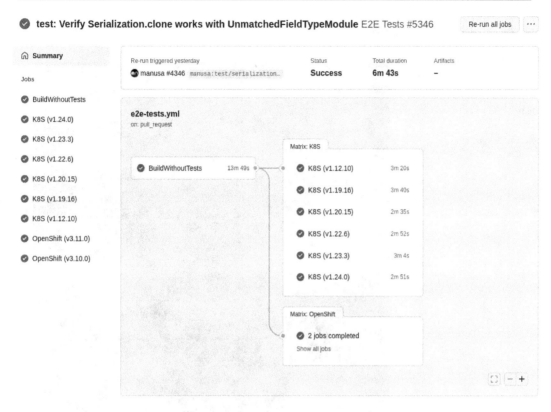

Figure 14.8 – A screenshot of a complex workflow run summary

From the workflow execution summary, we can also check out the details of each job by clicking on their descriptions. GitHub will redirect us to the job execution details section, where we'll be able to check the logs for each of the steps and manually re-run the job. Now, let's click on the **Build and test** job to see its details:

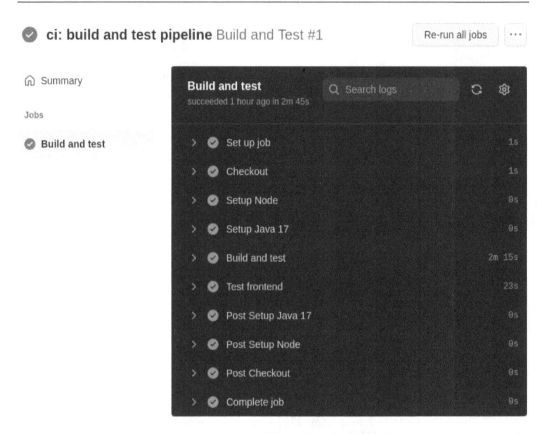

Figure 14.9 – A screenshot of the Build and test job's details

Notice how each step is identified by the name we provided in the YAML configuration file. Having explicit names that describe what each step does will help us quickly identify any problem whenever one of the steps fails. From this page, we can download the complete log for the job execution or expand each step to check their individual logs.

We've now created a CI pipeline that will make sure that any change we integrate into our main branch doesn't break the application. We can further prove this point by creating a pull request that introduces some code that breaks the project. You can try this yourself by creating a new branch with some broken code and creating a new pull request from the GitHub interface. I will try this by modifying the @ Path annotation of the TaskResource class; you can perform the same modification or change any other part of the application code (which has tests), maybe from the frontend. The following snippet shows my changes to the TaskResource class:

```
@Path("/api/v1/tasks-broken")
```

After pushing my changes to a new branch, and creating a pull request, a new workflow execution should start and complete with failures. These failures should be visible in the pull request interface:

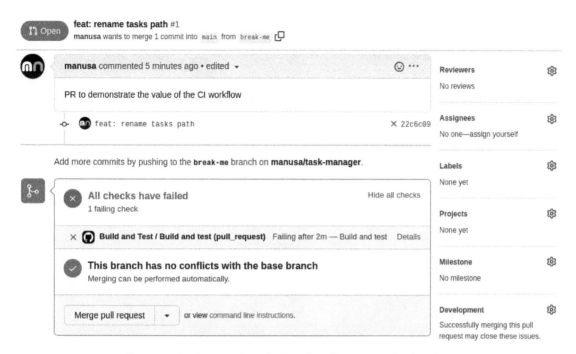

Figure 14.10 – A screenshot of a GitHub pull request with a failed job

This example clearly illustrates how we can use GitHub Actions and its tight integration with the rest of the GitHub interfaces to adopt a successful CI development practice. GitHub even allows us to define some required checks to be able to merge a pull request, so we could enforce a successful build before merging any changes into the main branch. This would be a good first step before moving into more complex development workflows, such as continuous delivery or deployment.

Summary

In this chapter, we learned how to create a CI pipeline for our task manager application with GitHub Actions. We started by learning about continuous integration, delivery, and deployment, and how these concepts are related to GitHub Actions. Then, we learned how to create a GitHub repository for our project, and how to push the application files we've been creating throughout this book. Finally, we learned how to create a GitHub Actions workflow to build and test our project to ensure future changes won't break the application and help us start our journey with CI/CD. You should now have a basic understanding of GitHub Actions and know how to implement pipelines.

Amazing job! You have reached the end of *Chapter 14, Creating a Continuous Integration Pipeline*, and the end of *Full Stack Quarkus and React*. You should now have enough skills and confidence to implement, test, and deploy full-stack web applications using Quarkus as a backend and React as the frontend on your own.

Questions

1. What is continuous delivery?
2. What is continuous deployment?
3. What is the advantage of GitHub Actions compared to other alternatives?
4. How do you initialize a local git repository?
5. Can you trigger a GitHub Actions workflow manually?

Appendix – Answers

Chapter 1

1. Quarkus is a Java web application framework similar to Spring Boot. Its main focus is on minimizing the application startup time and memory footprint while boosting the developer experience.

2. You can create a Quarkus project from scratch by leveraging its CLI tool, a Maven goal, or most conveniently, by using the `https://code.quarkus.io` page.

3. To run a Quarkus project in development mode, you need to invoke the following command: `./mvnw quarkus:dev`.

4. **TDD** stands for **test-driven development**, which is a software development process that aims to improve the developer's productivity and the overall code quality.

5. There are several packaging modes for Quarkus. You can easily package the application by running the following command: `./mvnw clean package`.

Chapter 2

1. Hibernate is one of the most mature **object-relational mapping** (**ORM**) libraries for the Java programming language. It also provides an implementation for the **Java Persistence API** (**JPA**) specification.

2. Yes, with Quarkus you can implement a fully reactive persistence layer.

3. You can implement an entity that leverages the Active Record pattern in Quarkus by extending the `PanacheEntity` base class.

4. An easy way to load initial application data to the database is by using the `quarkus.hibernate-orm.sql-load-script` configuration option and SQL scripts in the `resources` folder.

5. Quarkus Dev Services automatically provisions containers with the services our application relies on and configures them for our application, running in dev mode or its integration testing phase.

Chapter 3

1. Yes, RESTEasy Reactive can be used to implement blocking endpoints too.

2. The two types provided by Mutiny to start a reactive pipeline are `Uni` and `Multi`.

3.   A Bean is an object managed by a CDI container that supports a set of services such as life cycle callbacks, dependency injection, and interceptors.

4.   Annotating a class with the `@ApplicationScoped` annotation is the easiest way to declare a singleton bean in Quarkus.

5.   bcrypt is the preferred password-hashing function because it's a slow algorithm, which makes it more resilient to brute-force attacks.

6.   To add a path parameter to a URL in Quarkus, you can declare it in the path by enclosing it in curly braces and then use the `@PathParam` annotation to retrieve it.

7.   It's not necessary to add an `@Produces` annotation if the response type is JSON and the `quarkus-resteasy-reactive-jackson` dependency is declared, since it will be automatically inferred.

8.   You can intercept Java exceptions to map them to HTTP responses by implementing `ExceptionMapper`

Chapter 4

1.   **JWT** stands for **JSON web token**, a proposed standard that can be used to represent and securely exchange claims between two parties.

2.   To verify a JWT signature, we'll need the public key.

3.   To generate a JWT in Quarkus, you can use the SmallRye JWT build dependency.

4.   Yes, we need to store a copy of the configured keys; however, the configuration for the path where these keys are stored can be overridden at runtime.

5.   We can use the `@ConfigProperty` annotation to retrieve a configuration value in Quarkus.

6.   If the `@RolesAllowed` annotation is applied both at a class level and then in a specific method, the method annotation takes precedence over the other.

Chapter 5

1.   We should use the `%test.quarkus.hibernate-orm.sql-load-script` configuration property and SQL scripts in the `resources` folder to add data to the testing environment database.

2.   You can automatically create a test suite class in IntelliJ by right-clicking on a production class declaration and clicking on the **Go To Test** menu.

3.   When you run a Quarkus test, Quarkus starts your application in the background.

4.   The `@TestSecurity` annotation overrides the application's security context when the test is run.

Chapter 6

1. GraalVM Native Image is an ahead-of-time compilation technology that generates standalone native executables.

2. Quarkus moves to the build phase many of the tasks that regular Java applications perform at runtime, which reduces drastically its startup time.

3. No, native image packaging doesn't always improve the application's performance.

4. Short-lived applications benefit from native image packaging.

5. The `quarkus.native.resources.includes config` property can be used to include additional resources in the application's native image.

6. Because we'll be distributing our application as a Linux container image.

Chapter 7

1. Material Design is a system of guidelines and tools created by Google to support the best practices and principles of good design when creating and designing user interfaces.

2. A frontend routing library can be useful to create bookmarkable and user-friendly URLs for your application.

3. No, React doesn't prescribe any routing library.

4. Redux Toolkit is an opinionated distribution of Redux, which includes utilities that simplify its configuration for the most common use cases.

5. React hooks allow users to use state and other features without writing a class.

6. **JSX** or **JavaScript XML** is an extension to the JavaScript language provided by React.

7. React guidelines recommend composition.

Chapter 8

1. Cross-origin resource sharing is a mechanism that allows a server to share resources to origins different from its own.

2. The easiest way to consume an HTTP API and persist its results in a Redux store is by using the `createApi` function provided by Redux Toolkit Query.

3. Redux Toolkit Query supports read-only `query` operations and read and write `mutation` operations.

4. In Material Design, Snackbars are used to display temporary short messages near the bottom of the screen.

5. The default URL for React's development server is `http://localhost:3000`.

Chapter 9

1. The most appropriate MUI component to display short messages in a way that attracts the user's attention is `Alert`.

2. React's built-in `useState` hook is used to preserve a local state within a component.

3. The action buttons in a modal dialog are usually located at the bottom.

4. You need to add the reducer and its middleware to the main application's store.

5. You need to configure the route path with a `param` definition, which can then be retrieved using the `useParams` hook.

Chapter 10

1. The `describe` function blocks are used to group test cases that are related to each other.

2. Jest will execute the `beforeAll` function once before any of the tests in the surrounding `describe` block are executed.

3. Yes, you can run asynchronous tests in Jest.

4. You can use the `userEvent` function to emulate a user typing into a text field.

5. With the Create React App test script and Jest, calculating the code coverage is as easy as specifying the `--coverage` argument when executing the `npm test` command.

Chapter 11

1. Microservices scale easily, provide high redundancy, and are programming language-agnostic.

2. Microservices require additional infrastructure, are harder to debug, and are more challenging when dealing with data consistency and transactionality.

3. You can configure the Maven Resources Plugin to include resources from multiple locations by using the `resources` configuration and adding an entry for each location.

4. The Maven Exec plugin allows the execution of programs and Java programs in separate processes.

5. To include additional resources in a native image, you can use the `quarkus.native.resources.includes` configuration property or the `resources-config.json` configuration file.

Chapter 12

1. Kubernetes manages container-based applications.

2. Container-based applications are those that are distributed using container images, which allow system administrators to treat applications as standard units that can be managed in a uniform and consistent way.

3. Application containers provide environment consistency, lightweight runtimes, efficient resource consumption, and many more advantages.

4. A Kubernetes Pod is an object that represents the smallest deployable computing unit that can be managed in Kubernetes.

5. A Kubernetes Ingress is an object to configure external access and routing to a Service deployed in the cluster.

6. No, Jib doesn't require a Docker-compatible daemon.

Chapter 13

1. Fly.io allows developers to deploy their workloads close to where their users are, improving the latency and performance of their applications.

2. You can override the configured database URL when running a Quarkus native application by providing the `-Dquarkus.datasource.reactive.url` command-line argument.

3. Yes, JKube supports Dockerfiles too.

4. `flyctl` reads the application's deployment configuration from the `fly.toml` file.

5. Yes, you can deploy existing container images to Fly.io.

Chapter 14

1. Continuous delivery is a natural extension of continuous integration, where each set of changes is not only integrated into a central repository but is ready to be deployed into production after a successful build and test iteration, in a safe, quick, agile, and sustainable way.

2. Continuous deployment is an extension of continuous delivery and includes the release and deployment of the application steps to the end of the automated pipeline.

3. GitHub Actions is tightly integrated into the rest of the GitHub ecosystem.

4. You can initialize a local Git repository by executing the following command: `git init -b main`.

5. Yes, you can trigger a GitHub Actions workflow manually, provided that you configure it using the `workflow_dispatch` event.

Index

Packt.com

Subscribe to our online digital library for full access to over 7,000 books and videos, as well as industry leading tools to help you plan your personal development and advance your career. For more information, please visit our website.

Why subscribe?

- Spend less time learning and more time coding with practical eBooks and Videos from over 4,000 industry professionals

- Improve your learning with Skill Plans built especially for you

- Get a free eBook or video every month

- Fully searchable for easy access to vital information

- Copy and paste, print, and bookmark content

Did you know that Packt offers eBook versions of every book published, with PDF and ePub files available? You can upgrade to the eBook version at packt.com and as a print book customer, you are entitled to a discount on the eBook copy. Get in touch with us at customercare@packtpub.com for more details.

At www.packt.com, you can also read a collection of free technical articles, sign up for a range of free newsletters, and receive exclusive discounts and offers on Packt books and eBooks.

Other Books You May Enjoy

If you enjoyed this book, you may be interested in these other books by Packt:

Full Stack Development with Spring Boot and React

Juha Hinkula

ISBN: 9781801816786

- Make fast and RESTful web services powered by Spring Data REST.
- Create and manage databases using ORM, JPA, Hibernate, and more.
- Explore the use of unit tests and JWTs with Spring Security.
- Employ React Hooks, props, states, and more to create your frontend.

Full Stack FastAPI, React, and MongoDB

Marko Aleksendrić

ISBN: 9781803231822

- Discover the flexibility of the FARM stack

- Implement complete JWT authentication with FastAPI

- Explore the various Python drivers for MongoDB

- Discover the problems that React libraries solve

- Build simple and medium web applications with the FARM stack

Packt is searching for authors like you

If you're interested in becoming an author for Packt, please visit `authors.packtpub.com` and apply today. We have worked with thousands of developers and tech professionals, just like you, to help them share their insight with the global tech community. You can make a general application, apply for a specific hot topic that we are recruiting an author for, or submit your own idea.

Hi!

I am Marc Nuri San Felix, author of Full Stack Quarkus and React. I really hope you enjoyed reading this book and found it useful for increasing your productivity and efficiency in Quarkus and React.

It would really help me (and other potential readers!) if you could leave a review on Amazon sharing your thoughts on Full Stack Quarkus and React.

Go to the link below or scan the QR code to leave your review:

`https://packt.link/r/180056273X`

Your review will help me to understand what's worked well in this book, and what could be improved upon for future editions, so it really is appreciated.

Best Wishes,

Download a free PDF copy of this book

Thanks for purchasing this book!

Do you like to read on the go but are unable to carry your print books everywhere?

Is your eBook purchase not compatible with the device of your choice?

Don't worry, now with every Packt book you get a DRM-free PDF version of that book at no cost.

Read anywhere, any place, on any device. Search, copy, and paste code from your favorite technical books directly into your application.

The perks don't stop there, you can get exclusive access to discounts, newsletters, and great free content in your inbox daily!

Follow these simple steps to get the benefits:

1. Scan the QR code or visit the link below:

https://packt.link/free-ebook/9781800562738

2. Submit your proof of purchase
3. That's it! We'll send your free PDF and other benefits to your email directly